A SHEARWATER BOOK

The New
Agrarianism

The New Agrarianism

Land, Culture, and the Community of Life

Eric T. Freyfogle

Island Press / SHEARWATER BOOKS
Washington • Covelo • London

A Shearwater Book
Published by Island Press

Copyright © 2001 by Island Press

Shearwater Books is a trademark of The Center for Resource Economics.

Library of Congress Cataloging-in-Publication Data
The new agrarianism : land, culture, and the community of life / Eric T. Freyfogle, editor.
    p.   cm.
Includes bibliographical references (p.   ).
    ISBN 1-55963-920-2 (cloth : alk. paper) — ISBN 1-55963-921-0 (paper : alk. paper)
    1. Land tenure—Environmental aspects—United States. 2. Land use—Environmental aspects—United States. 3. Sustainable agriculture—United States. 4. Conservation and natural resources—United States. 5. Human ecology—United States. I. Freyfogle, Eric T.
    HD205 .N49 2001
    333.73—dc21   2001004025

British Cataloguing-in-Publication data available.

Printed on recycled, acid-free paper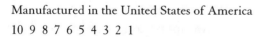

Manufactured in the United States of America
10 9 8 7 6 5 4 3 2 1

For Mercy and Bill Davison,
Julianne Lutz Newton, and Todd Wildermuth

Who is the land? We are, but no less the meanest flower that blows.

—Aldo Leopold

For nitrates are not the land, nor phosphates; and the length of fiber in the cotton is not the land. Carbon is not the man, nor salt nor water nor calcium. He is all these, but he is more, much more; and the land is much more than its analysis.

—John Steinbeck

There is no way to separate feeling from knowledge. There is no way to separate object from subject. There is no good way and no good reason to separate mind or body from its ecological and emotional context.

—David Orr

We will always be rewarded if we give the land credit for more than we imagine, and if we imagine it as being more complex even than language.

—Barry Lopez

The defence and salvation of the body by daily bread is still a study, a religion, and a desire.

—Thomas Hardy

The higher processes of art are all processes of simplification.

—Willa Cather

Consider the lilies of the field, how they grow. . . .
—Matthew 6:28

It seems wiser to be moderate in our expectations of nature, and respectful; and out of so simple a thing as respect for the physical earth and its teeming life comes a primary joy, which is an inexhaustible source of arts and religions and philosophies.

—John Crowe Ransom

Blessed is the man who loves the soil!

—John Burroughs

It has never been our national goal to become native to this place. It has never seemed necessary even to begin such a journey. And now, almost too late, we perceive its necessity.

—Wes Jackson

It is the most arrant presentism to say that a philosophy cannot be practiced because the philosophy is found in the past and the past is now gone.

—Richard Weaver

We can lie to ourselves about many things, but if we lie about our relationship to the land, the land itself will suffer, and soon we and all other creatures that share the land will suffer.

—Scott Russell Sanders

Where is our comfort but in the free, uninvolved, finally mysterious beauty and grace of this world that we did not make, that has no price? Where is our sanity but there? Where is our pleasure but in working and resting kindly in the presence of this world?

—Wendell Berry

Contents

Introduction

A Durable Scale

With no fanfare, and indeed with hardly much public notice, agrarianism is again on the rise. In small corners and pockets, in ways for the most part unobtrusive, people are reinvigorating their ties to the land, both in their practical modes of living and in the ways they think about themselves, their communities, and the good life. Agrarianism, broadly conceived, reaches beyond food production and rural living to include a wide constellation of ideas, loyalties, sentiments, and hopes. It is a temperament and a moral orientation as well as a suite of economic practices, all arising out of the insistent truth that people everywhere are part of the land community, just as dependent as other life on the land's fertility and just as shaped by its mysteries and possibilities. *Agrarian* comes from the Latin word *agrarius,* "pertaining to land," and it is the land—as place, home, and living community—that anchors the agrarian scale of values.

For contemporary adherents, in cities as well as rural areas, agrarian traditions have provided a diverse set of tools for fashioning more satisfying modes of life. And as the writings here reveal, they are making extensive use of those tools, to strengthen families and local communities, to shape critiques of modern culture, and in various ways and settings to mold their lives to their chosen natural homes.

I

As a collection of practices and principles, agrarian*ism* has enjoyed a long and curious history in recorded Western life, from ancient Greece to the present.[1] Prominent in that history, of course, have been the methods and economies of gaining food from fields, forests, and waters. Just as important, though, have been the ways that farm life has figured in a people's social and moral imagination. Agrarianism's central image has long been (to use Southern writer Andrew Lytle's term) the livelihood farm—the well-run farmstead that provides the locus and cultural center of a family's life, the place where the young are socialized and taught, where stories arise and are passed down, where leisure is enjoyed, where the tasks of daily living are performed, and where various economic enterprises take place, in garden, orchard, kitchen, woodlot, toolshed, and yard.

Such a farmstead, diverse in crops and livestock, has stood in the agrarian imagination as a model incubator of virtue and healthy families. It has exemplified the traditions and possibilities of essential work, well done, in familiar settings. It has linked humankind to other forms of life, to soil and to rains, and to cycles of birth, death, decay, and rebirth. In its independence it has provided both a haven from corrosive cultural values and much-needed ballast to stabilize civil states. Generation upon generation, people have retreated to such farms in times of strife, figuratively if not literally, in order to heal, regroup, and set out anew.

Given this history, it is as unsurprising as it is heartening that agrarian ways and virtues are resurging in American culture, prompted by a wide range of public and private ills. To the diseases and degradations of the modern age, a New Agrarianism is quietly rising to offer remedies and defenses, not just to the noise, vulgarity, and congestion that have long affronted urban dwellers but to the various assaults on land, family, religious sensibilities, and communal life that have tended everywhere to breed alienation and despair.

Evidence of the New Agrarianism appears today all across the country, in the lives and work of individuals, families, and community groups:

- In the community-supported agriculture group that links local food buyers and food growers into a partnership, one that sustains farmers economically, promotes ecologically sound farm practices, and gives city dwellers a known source of wholesome food

- In the woodlot owner who develops a sustainable harvesting plan for his timber, aiding the local economy while maintaining a biologically diverse forest

- In the citizen-led, locally based watershed restoration effort that promotes land uses consistent with a river's overall health and beauty

- In the individual family, rural or suburban, that meets its food needs largely through gardens and orchards, on its own land or on shared neighborhood plots, attempting always to aid wildlife and enhance the soil

- In the citizen-led effort to promote greenbelts and recreational trails designed not just for human use but also to mitigate storm water runoff, improve wildlife habitat, and promote compact housing developments

- In the farmer who radically reduces a farm's chemical use, cuts back subsurface drainage, diversifies crops and rotations, and carefully tailors farm practices to suit the land

- In the family—urban, suburban, or rural—that embraces new modes of living to reduce its overall consumption, to integrate its work and leisure in harmonious ways, and to add substance to its ties with neighbors

- In the artist who helps community residents connect aesthetically to surrounding lands

- In the native plant and game bird societies that promote locally tailored landscape plans to aid biodiversity

- In the faith-driven religious group that takes seriously, in practical ways, its duty to nourish and care for its natural inheritance

- In the motivated citizens everywhere who, alone and in concert, work to build stable, sustainable urban neighborhoods; to repair blighted ditches; to stimulate government practices that conserve lands and enhance lives; and in dozens of other ways to translate agrarian values into daily life

Many worries and hopes lie behind this welling up of interest in land-centered practices and virtues. The degradation of nature—problems such as water pollution, soil loss, resource consumption, and the radical disruption of plant and wildlife populations—is everywhere a core concern. Other worries center on food—its nutritional value, safety, freshness, and taste—and on the radical disconnection today, in miles and knowledge, between typical citizens and their sources of sustenance. Then there are the broader anxieties, vaguely understood yet powerfully felt by many, about the declining sense of community; blighted landscapes; the separation of work and leisure; the shoddiness of mass-produced goods; the heightened sense of rootlessness and anxiety; the decline of the household economy; the fragmentation of families, neighborhoods, and communities; and the simple lack of fresh air, physical exercise, and the satisfactions of honest, useful work. Permeating these overlapping concerns is a gnawing dissatisfaction with core aspects of modern culture, particularly the hedonistic, self-centered values and perspectives that now wield such power.

The writings gathered here present various perspectives on this New Agrarianism. All date from 1986 and later, and all were chosen for their literary merit. Some selections use the term *agrarian* expressly; others leave their alternative way unnamed. Some explore the virtues and possibilities of agrarian practices; others dissect the ills and allures of the dominant American culture. Although most selections are nonfiction, two of them present agrarian perspectives in fictional settings. Whether fiction or non-fiction, most of the writings do not assess agrarianism abstractly so much as illustrate it in action. And it is appropriate that they do so, given that agrarianism is about ideas and principles chiefly as they inspire and are revealed in the lives, values, and hopes of adherents. One learns the agrarian way best by observing it in action, by see-ing how agrarians orient their lives toward land and how, from that base, they look outward to critique the surrounding world.

As the readings here illustrate, the New Agrarianism of the past generation has pruned key elements from older agrarian ways while nourishing other shoots and stimulating new ones. Gone entirely is the old slave-based, plantation strand of agrarianism; a regional variant to begin with, it deviated markedly from the family-based homestead ideal. Still around but much cut back are the once powerful assumptions about gender roles within the fam-ily and the larger household economy: As much as other Ameri-cans, agrarians have struggled to promote fairness and individual choice without losing the benefits of specialized labor. On the flour-ishing side, there is the heightened interest today in land conserva-tion, which has taken on a distinctly ecological cast. Much strength-ened, too, has been the New Agrarian challenge to materialism and to the dominance of the market in so many aspects of life. And yet, even with its new shapes and manifestations, agrarianism today remains as centered as ever on its core concerns: the land, natural fertility, healthy families, and the maintenance of durable links between people and place.

Agrarianism is very much alive and flourishing in America today, in ways both new and old and in diverse vocations and avo-cations. One could not call it a major element of contemporary cul-ture, yet once aware of agrarianism, one stumbles on its outcrop-pings at many a turn. Within the conservation movement, the New Agrarianism offers useful guiding images of humans living and

working on land in ways that can last. In related reform movements, it can supply ideas to help rebuild communities and foster
greater virtue. In all settings, agrarian practices can stimulate hope
for more joyful living, healthier families, and more contented, centered lives.

II

Agrarians have typically been happier to live their lives than to
write about them. Reports on agrarian ways tend toward the frag-
mentary and the narrative, covering bits and pieces, less often ana-
lyzing or proceeding by dialectic than illustrating and evoking.
Summaries of agrarianism do exist, but they tend to define the
agrarian way too narrowly.[2] Thus, one finds summaries confining
agrarianism to food-related economic practices, insisting that the
"land" in agrarianism means only farm fields, concluding that
agrarianism is merely a cloaked special-interest demand by farmers
for a greater slice of the economic pie, or patronizing it as nothing
more than a Currier and Ives–style retreat from the stringencies of
modernity. A faithful characterization needs to cast its net more
widely and fairly. It also must remain sensitive to change over time,
for agrarianism is very much a living as well as a lived tradition.

What, then, are the principal elements and themes of the New
Agrarianism that emerge from its many writings and manifestations?

The place to begin, naturally, is with the agrarian root—the
land itself and how it is conceived. For agrarians, land is an organic
whole, teeming, when well tended, with an abundance of plant and
animal members. Humans are special members of that living community, but they are members nonetheless, not onlookers from
afar: They are as linked and embedded as the land's many other
creatures. In embracing this view, agrarians reject the conceit that
the land is merely a warehouse of discrete natural resources. They
reject, too, the claim that humans are or can be autonomous in rela-

tion to the natural places they inhabit. Land may not be the source of all wealth, as eighteenth-century Physiocrats claimed,[3] but it remains the essential base of all terrestrial life. In every reading in this volume, the land looms as a major presence.

From this recognition of interconnected life comes an overriding attentiveness to the health of the land. In the agrarian mind, the health of humans is dependent in the long run on the well-being of the larger land community. English reformer Sir Albert Howard summed up the point a half-century ago in his work *The Soil and Health* when he urged readers to understand "the whole problem of health in soil, plant, and animal, and man as one great subject."[4] This holistic idea also guided the work of conservationist Aldo Leopold, who was as responsible as any person for bringing ecology to bear on agrarian concerns. The overall well-being of the land community, its "integrity, stability, and beauty," became the focal point of Leopold's influential land ethic.[5] Among contemporary writers, Wendell Berry has been particularly forceful in drawing attention to the health of the natural whole, to "the one value," the one "absolute good," that undergirds our agnostic culture. In its fullest sense, Berry argues, health makes sense only when defined at the land-community level; such a community "is the smallest unit of health." "To speak of the health of an isolated individual," in Berry's view, "is a contradiction in terms."[6] The first of Berry's two essays here, "The Whole Horse," presents with particular clarity the agrarian preoccupation with the land's lasting vigor.

Guided by their organic perspective, agrarians pay close attention to the way people in their daily lives interact with particular lands, near and far, directly and indirectly. The product cycle looms especially large in this understanding: where raw materials come from and how they are produced, particularly food, fiber, and energy, and where wastes go and with what effects on which communities. "Nothing arises but from death," Lucretius observed long ago, and it is with constant awareness of this reality that agrarians comprehend their life patterns in cyclical terms. The wheel of life is no mere metaphysical ideal; it is an apt description of how the land's fertility is maintained as plants and animals die and nourish the soil, which in due course yields new life. To the agrarian, the soil is the great terrestrial connector of life, death, and new life, the very medium of resurrection. Indeed, so important is the soil and its fertility that agrarians are sometimes accused of soil worship by those less impressed by its vital, creative role.[7] Agrarian writer Paul B. Thompson frames this dependence as follows:

> Farming's essence is true to soil. Proper farming might be said to make concrete what is latent in humanity's dependence upon the earth, for the act of good farming both releases and replenishes the provisions for human sustenance. Farming is the activity that locates the human species most surely in the planetary ecosystem of the earth. It is on farming that we depend for food, and in farming that what we take from the earth is returned to it.[8]

Among the writings here, the selections by Alan Thein Durning and David W. Orr deal particularly with the product cycle. In the work of Wes Jackson, so vividly portrayed by Scott Russell Sanders in "Learning from the Prairie," one sees the overriding importance of fertile soil.

The product cycle, from earth to consumer good to waste, traces not just lines of dependence and causation but also lines of responsibility. Dissenting from the modern view, agrarians believe that those who buy products are implicated morally in their production, just as those who discard waste items are morally involved in their final end. Those who hire a trash hauler to take garbage away are not cleansed of their complicity in its disposal, any more than buyers of chemically bathed apples are insulated from the ills of orchard management. Producers and sellers, too, are morally responsible for their work, and in ways the market cannot absolve or cleanse when their products are sold.

This assignment of complicity is part of the larger recognition by agrarians that membership in a land community necessarily entails responsibilities, chiefly to the community as such. One cannot live in a place without altering it, yet alterations differ vastly in their effects on the health of the land. The agrarian aim is not to minimize effects on nature, as if human change were necessarily evil. It is to harmonize them: to craft ways of living in a place that are respectful of the land's long-term fertility and that accommodate, insofar as possible, the human penchant to err and make messes.[9]

Much of agrarian culture has to do with the particulars of these responsibilities, with making the translation in daily life from abstract senses of membership and duty to particular patterns of living. Although the science of ecology now increasingly informs these issues, the challenge at root is an ethical one, dealing as it does with the rightful human role in the order of Creation. Right living on the land is infused with moral dimensions, and sustaining land health is a moral guideline if, indeed, not a moral imperative. Given this moral center, agrarianism stands in contrast to the moral relativity of the modern day, the pernicious illusion that one set of values is as good as another. Agrarianism embraces a responsible form of individualism, what social critic Richard Weaver years ago termed social-bond individualism, as opposed to the anarchic individualism (Weaver's term) or bogus individualism (Leopold's term) that lies behind libertarian calls for maximum freedom and minimal responsibility.[10] Agrarianism, then, sees hope in the modest resurgence in America of interest in public virtues and moral discourse, insisting only and emphatically that virtue prevail in all aspects of life, not just within the family but also on the job, on the land, in corporate boardrooms, and in legislative chambers.

The infusion of moral concerns into all aspects of life is a natural offshoot of the agrarian's unwillingness to fragment the human condition. Here, the farmstead provides a continuing reminder. Work and leisure, the secular and the sacred, the functional and the beautiful, all retain an elemental integrity. Life is not starkly divided between work, school, and home; between production and consumption; between means and ends. These themes emerge vividly on the pages that follow in the real-life story of the Kline family in Ohio and in Wendell Berry's masterful tale "The Boundary."

III

Good land use—perhaps the highest agrarian aspiration—is by no means an easy undertaking, as agrarians well know. In demanding it of themselves and one another, they recognize the difficulty of the task they have set.

Good land use requires an intimate knowledge of land together with high levels of skill. Farming in particular is as much an art as a science, given the vagaries of nature and the inadequacies of the human mind. Then there is the whole matter of economics and the recognition that sustainable land use is a practical ideal only when financially feasible. To identify these realities is to set forth the prime challenges to which agrarian proposals respond.

As agrarians see things, good land use over time depends on a local culture that is durable and economically successful. Such a culture necessarily crosses generations, and it is sustained, as Wendell Berry has emphasized, by a handing down of wisdom within the local land-using community from neighbor to neighbor and generation to generation.[11] Although book learning and scientific studies are important, good land use requires the tailoring of general precepts to particular land parcels, work that can be done only by a person attentive to a parcel and committed to its long-term fertility. Long-term perspectives arise most readily when owners feel committed to the lands they own, when they view them less as economic assets—and hardly at all as market commodities—than as homes, livelihoods, and treasures, tended by one generation and passed along in time to the next.

Good labor on the land means working with nature, attending

to its possibilities, respecting its mysteries, and remaining alert to its penchant to surprise. Good work, agrarians recognize, often takes time, and some jobs cannot safely or wisely be speeded up. Bad work, on the other hand—bad in terms of adverse effects on the land community and the social order—can happen quickly and leave enduring scars in its wake. In the stock pastoral tale, the fictional hero escapes from a corrupted city and flees to a pristine, wholesome wild, there to begin life in a new Eden. Agrarian writers of recent decades have had a far different story to recount. Not Eden but a battle-weary land commonly greets the agrarian pilgrim today, a land marred by eroded hills, polluted rivers, and biologically impoverished forests.

Success in such a challenging life necessarily depends on a constant and careful attentiveness to the land. Each land parcel is unique, to cite a bedrock agrarian adage, which means that good land use necessarily varies from place to place. To work with the land responsibly is to converse with it in a type of dialectical interchange. Such a conversation begins, among the best of agrarians, with close attention to what nature would do in a place when left alone. What does nature have to offer in a given place? What will nature allow human users to do? What will it help them do?[12]

At the base of agrarian thought about land use is the fundamental recognition that nature is far bigger than humans, bigger than they know or even can know. Human knowledge of nature is limited, encased within layers of mystery. To base land-use decisions solely on empirical data is to invite disaster, given the vast gaps in what even well-skilled humans understand. Good land use requires a mixing of the empirical and rational with the intuitive and sentimental. Embedded within nature are whole realms of wisdom that humans have hardly noticed, much less mastered. "Nature as measure," a phrase first offered by Wendell Berry, has become a widely used agrarian proverb.[13] Good land use everywhere is undertaken with humility, in a type of trial-and-error or conversational interaction that respects nature as a wise and full partner. As Scott Russell Sanders' portrait here shows, the staff at The Land Institute has translated these truths into a refined and productive art.

Because good land use often results in lower short-term yields, agrarians are painfully aware of economic realities. No land use can endure if it makes no economic sense, and in the short run at least, good land use is more costly than bad land use. In the short run, plowing hillsides raises yields while eroding soil. Inorganic fertilizers, chemical pesticides, fossil fuels, expansive monocultures, and extensive subsurface drainage all cut costs on the farm while sapping overall land health. Predictably, agrarians are sharp critics of cheap-food policies that push landowners to abuse land by cutting corners. They condemn, too, free-trade policies that pit landowners in one part of the globe against landowners in another, policies that in practice test not so much who is the most efficient (although that factor plays in) as whose lands are most naturally endowed and whose land ethics are lowest. Agrarians fight back by promoting collective agrarian efforts, recognizing that the individual landowner alone has little chance of effecting change. For more than a century, cooperative buying and marketing efforts have been a staple agrarian response. In recent years, agrarians have sought local outlets for their produce, particularly specialty outlets that pay premiums for fresh, chemical-free crops. If production controls appeared more feasible, agrarians might support such measures, too, as a way to reduce destructive competition and assure farmers of sufficient income to allow them to use the land well.

The agrarian concern for economic stability and durability accounts in part for agrarians' insistence that the household remain what it typically was in the United States until well into the twentieth century—a center of economic production, meeting its needs from within so as to reduce dependence on the market. Agrarians also foster market independence by producing multiple crops rather than a single market staple; by lowering purchased inputs insofar as possible; and, when feasible, by resorting to barter or cash substitutes. Independence not possible within a single household can increase within a neighborhood of like-minded agrarian households through exchanges and sharing. What is good for the neighborhood is, to the agrarian mind, good for a larger community as well. A community, too, should reduce its dependence on the outside market by adding value to outgoing products, by purchasing needed materials as raw goods rather than as fully processed ones, and by fostering internal economic diversity. All these measures, of course, cut against the free-traders' ardent call for specialization and interdependence. But free-traders, as agrarians painfully know, care little about the health of particular households and communities, just as in their quest to lower market prices they discount ecological scars.[14] Conflict is inescapable.

IV

Agrarians hold tenaciously to the institution of private property, viewing it as part of the bedrock on which their world rests. Governments, they agree, should own lands that serve distinctly public functions, including ecological ones. But lands devoted to production are typically best used under the responsible control of private owners.

In the agrarian mind, however, private property has a much more restrained meaning than it does to the finance capitalist. Property ownership is a core civil right only when property use is linked to and reflects the personality of a particular morally responsible owner. Both in word and in theory, *property* is linked to such related terms as *propriety, proper,* and *appropriate*.[15] To own property is to make it one's own, to mix one's labor with it and attach it to one's moral domain. Only such an owner can envelop land with long-term dreams and link it to the maintenance and reputation of a family. For this reason, not just government ownership but also widespread tenancy arrangements worry the agrarian mind.[16] Neither a landlord nor a tenant has the same perspective on land as does an owner-user. Moreover, property claims are weakened when an owner does not live on the land owned or owns too much land to know it intimately and use it carefully. One strand of agrarian thought, still alive today, would distinguish between family property and investment or capitalist property, reserving full legal protection of property rights only for the former.[17] As for the widespread opposing view of private property as unbridled license to exploit, agrarians today are inclined to respond with the kind of scorn William Kittredge evinces here in "Owning It All."

In particular regions, local agrarian traditions honor the sharing or joint management of working lands, hearkening to common-field farming methods of medieval England, though one rarely encounters in the United States instances of true commons management of the type still alive elsewhere in the world. Shared-management arrangements are by no means viewed as inconsistent with the core values of property ownership. Indeed, as ecological concerns weigh more heavily, agrarians are seeing merit in the once familiar wisdom that shared governance of neighboring lands and resources can sometimes be better adapted to nature's limits, most evidently in the case of arid grazing lands and common-pool resources such as water flows and fisheries. Encouraged by conservation organizations, agrarians are also seeing greater need to talk about their respective land-use practices at the ecosystem or watershed scale, identifying shared problems and considering what they might reasonably expect from one another in terms of fostering the land's health. On this issue, one detects a distinct generational shift in the agrarian mind: Younger agrarians appear more willing than older colleagues to identify needs for joint action and to see benefits in formal land-planning processes. Brian Donahue probes the attractions and possibilities of shared land management here in his "Reclaiming the Commons."

Agrarians for various reasons have been slow to admit any need for formal land-use regulations. As much as other Americans, they dislike being told what to do. Believing they know their lands better than others do, they resent the claimed expertise of outsiders; fearing for their economic independence, they worry that regulations intended to protect the land might unintentionally push them to abuse it. And yet the core aims of land-use laws resonate with the agrarian mind, particularly laws to conserve soil, clean water, and human health. In addition, the idea that landowners owe duties to the community is one that draws ready assent. Although sympathetic with libertarian rhetoric, true agrarians ultimately reject it because it provides too much of a shield for destructive practices and unfair competition. Private land-use decisions deserve protection only when they are respectful of neighbors and reasonably consistent with the well-being of the overall land community.

Despite their embrace of the community's moral claims, agrarians have generally been reluctant to transform moral duties into binding legal ones.[18] Land-use laws in particular are often suspect, given their perceived use by outsiders as tools to disrupt and restrict local life. Rural agrarians have been especially slow to see how a local community can use land-use laws to defend its interests, both against invasive outside forces and against the invidious practices of community residents who fail to do their part in sustaining the common good. In reality, far more land-use regulation is likely to be necessary if agrarian practices and virtues are to survive in a day of accelerating urban sprawl and profit-driven manipulations of genetic codes.[19] Outside the land-use realm, individual agrarians can successfully resist outside pressures they dislike—by home schooling and religious worship, for instance. But there are many other pressures, economic and land-related ones in particular, that can easily crush an unorganized people. A community that views its topsoil as a shared treasure needs to act when landowners are destroying it; a community that views its local river as a lifeline needs to set limits on how much landowners can pollute it and disrupt its flow. As communitarian theorist Amitai Etzioni observes, "law in a good society is first and foremost the continuation of morality by other means."[20] Regulation is often a necessary tool to defend the community against violations of communal standards.

Older forms of agrarianism often had, as a defining plank in their platforms, calls for widespread, equitable division of land.[21] The idea was endorsed by Thomas Jefferson and J. Hector St. John de Crèvecoeur in the late eighteenth century, and it undergirded homestead laws and other land distribution schemes well into the twentieth century. For its proponents, dispersed ownership offered political benefits as well as economic ones: Landowners were reputedly more stable and responsible citizens, and more likely to support sound governance. Land distribution was also intended to help remedy urban congestion, low wages, and unemployment. For the most part, New Agrarians have dropped this issue. Few call for division of large farms into family-sized homesteads, for instance, or for new homestead laws that open up public lands to family settlement. There has been little talk even about limiting farmland ownership to farmers and their immediate families, a policy implemented in parts of northern Europe. Although agrarians would like to see far more people living close to the land, they recognize that good farming requires highly specialized talents and is hardly an activity that the uninitiated can take up full-time with any prospect of success. New Agrarians seek to foster a wide range of human–land connections, not just expanded private ownership. They hope that as many citizens as possible will become keenly aware of their dependence on land and will value highly its healthy functioning.[22] Many of the selections here explore ways of achieving these goals, particularly the essays by Dan Imhoff, Stephanie Mills, and Brian Donahue.

V

One cannot fully assess any holistic view of the human predicament without attending to its views of human nature. On this issue, too, the New Agrarianism needs situating on the intellectual spectrum.

European perspectives on human nature differed widely during the era of America's founding.[23] The tradition running through the thought of Niccolò Machiavelli, John Calvin, and Thomas Hobbes reflected a somber view of natural man and woman: Humans were flawed creatures, prone to evil, and much in need of control and correction by social, religious, and government structures. The alternative perspective, featuring such writers as John Locke and Jean-Jacques Rousseau, was far more sanguine. Humans were basically good and reasonable by nature: perhaps in need of social control to contain the few misfits (Locke); perhaps, though, so good that society was inevitably corrupting (Rousseau). By the early nineteenth century, America's culture largely embraced the more optimistic view, which flourished in the writings of Ralph Waldo Emerson and dominated the politics of Andrew Jackson. The darker view by no means disappeared, however; Nathaniel Hawthorne wrote of it and so did Herman Melville, and religious revivals brought regular reminders of human depravity and the need for God's grace.

Contemporary agrarians largely chart a middle course on this foundational issue: They trust humans to act well but recognize the need for communal order and moral codes to shape and guide human passions. The family, the neighborhood, multiple generations, a religious tradition—all are needed to help the individual live with dignity and grace. Civilization, in short, has a lot to do with the development of self-discipline. Agrarians would agree with Edmund Burke that humans are qualified for civil liberty in proportion to their willingness to put moral chains on their appetites. And they would largely agree with Amitai Etzioni's dynamic or developmental view: People are born savage but can become more virtuous through socialization, though perhaps "never as virtuous as individualists and welfare liberals envision."[24]

Despite their strong sense of individual responsibility, agrarians view humans chiefly as social beings, following in the long tradition associated with Aristotle. People thrive best when knitted into responsible community structures. Such structures, when they function well, can tolerate occasional misfits and contain their disruptive impulses. The social community, then, counts for a great deal in the agrarian scale, and civic-mindedness is a high virtue— as Scott Russell Sanders aptly displays in his second selection here, "The Common Life." As for the meaning of community, two related definitions would seem to capture the agrarian perspective. The first is Wendell Berry's:

> By community, I mean the commonwealth and common interests, commonly understood, of people living in a place and wishing to continue to do so. To put it another way, community is a locally understood interdependence of local people, local culture, local economy, and local nature.[25]

The second is Amitai Etzioni's:

> Community is defined by two characteristics: first, a web of affect-laden relationships among a group of individuals, relations that often crisscross and reinforce one another (rather than merely one-on-one or chainlike individual relationships), and second, a measure of commitment to a set of shared values, norms, and meanings, and a shared history and identity—in short, to a particular culture.[26]

Given their firm moral code and their mixed view of human nature, agrarians are prone to question radical egalitarianism, so influential in modern culture. Whether or not people start life equally, they soon spread apart, not just in their abilities and accomplishments but also in their self-discipline and moral temperaments. Agrarian families and communities embrace their weak members rather than turning them out. But they do so without confusing virtue with vice or foolishness with wisdom. Not equality but fraternity is the guiding precept in agrarian thought. Fraternity pays attention to differences in morals and behaviors, yet as Richard Weaver put it, as an "ancient feeling of brotherhood [it] carries obligations of which equality knows nothing."[27] As Wendell Berry observes, "if we have equality and nothing else—no compassion, no magnanimity, no courtesy, no sense of mutual obligation and dependence, no imagination—then power and wealth will have their way; brutality will rule."[28]

Agrarians value fraternal bonds all the more because they recognize that fate can be harsh and the rewards one obtains in life are often undeserved. Particularly among tillers of the soil, the mysteries and vagaries of nature undergird a tragic view of life. Working the land breeds a stolidity that spurns fads, grand hopes, and aggressive sales pitches.[29] Progress does occur, but agrarians have little faith in its inevitability. Change can be bad as well as good, and a good development overall can have bad elements within it. Many of the changes of the past century have brought not progress but its opposite or, rather, progress mixed with its opposite and overpowered by it. Still, agrarians push on, confident or at least hopeful that through good work long practiced, humans can better their plight.

One challenge of the New Agrarianism is to promote healthy families and communal structures without resorting to the unfair social constraints and discriminatory practices that have characterized the agrarian past. Agrarianism's history has not been uncheckered, any more than has America's history generally.[30] One cannot turn to it uncritically, that is, to find an innate set of embedded virtues. The New Agrarianism is out to rectify patriarchy and racism, yet it strives to do so without crushing communal structure in the process—without resorting to a radical individualism that, on the other side, can itself be pernicious.[31] The challenge for agrarians on this issue is the challenge faced by all citizens who value families, neighborhoods, and healthy communities: to promote virtue and responsible behavior without casting people in unfair and unyielding roles.

VI

Although limits, responsibilities, and worries infuse the agrarian way, to dwell on them is to distort the overall picture, for a flourishing agrarian life is first and foremost a life of positive joy. Nature gives rise to much of that joy, with all its splendors and surprises and even its occasional terrors. Just as important are the pleasures that come from exercising a variety of skills to meet basic needs. At its best—and its best, to be sure, is often not fully attainable—the agrarian life is an integrated whole, with work and leisure mixed together, undertaken under healthful conditions and surrounded by family. As best they can, agrarians spurn the grasping materialism of modern culture; they define themselves by who they are and where they live rather than by what they earn and own. Few writers have made these points more lucidly than Ohio farmer David Kline, featured here in "Great Possessions."

In the agrarian view, physical labor is by no means an evil to be avoided at all cost, although dull tasks repeated too long are by no means relished. Indeed, manual labor skillfully performed is part of the good life. On this point agrarians today carry forward a tradition begun among literary figures as early as Hesiod, continued by monastic leaders such as Saints Benedict and Bernard, and defended in the twentieth century by such eloquent spokespeople as John Crowe Ransom and homesteaders Helen and Scott Nearing. *Laborare est orare,* "Work is worship," was Thomas Carlyle's phrasing, so popular in the nineteenth century, and the religious overtones of good work remain alive among agrarians today. Conservative idealist Richard Weaver explained the role of good work in his classic study *Ideas Have Consequences:*

> Pride in craftsmanship is well explained by saying
> that to labor is to pray, for conscientious effort to real-
> ize an ideal is a kind of fidelity. The craftsman of old
> did not hurry, because the perfect takes no account of

time and shoddy work is a reproach to character. But character itself is an expression of self-control, which does not come of taking the easiest way. Where character forbids self-indulgence, transcendence still hovers around.[32]

Agrarians have long celebrated the amenities of life, hospitality, conversation (especially storytelling), and good manners without professing extraordinary skill in them. Such amenities center on the home, and one of the chief contentments of agrarians is the sense of having a distinct home, knowing that home, and feeling centered by it.

Home food production, an agrarian preoccupation everywhere, can be a particular source of pleasure, as generations of gardeners have known. Pleasure comes from exercising skill to meet basic needs and participating in the processes of natural growth. Pleasure also comes from knowing that food is fresh and uncontaminated. Unlike commercial growers (as Anne Mendelson illustrates here in "The Decline of the Apple"), agrarians select seed lines more for nutrition and taste than for external appearance or ease of shipping. They favor plants that produce over as long a season as possible, not ones whose yields ripen all at once. Food and other items produced by the household economy are rated chiefly for their use value within the household rather than for any exchange value. The true livelihood farm begins by addressing as many of its own needs as possible, turning to the market chiefly to meet needs that cannot be satisfied internally and to dispose of extra produce. Consistent with this focus, land is valued highly for its fertility and its ability to meet such needs, not just for the price it might fetch when sold. Fertility is best maintained through natural processes, and lands that flourish in their natural fertility are the most beautiful of places. In the agrarian aesthetic, the natural and the native rate high. Beauty is not a functionless decoration tacked on but rather an integral characteristic of appropriate, well-crafted elements of a working land.[33]

When all the pieces of the agrarian life come together—nutrition and health, beauty, leisure, manners and morals, satisfying labor, economic security, family and neighbors, and a spiritual peacefulness—we have what agrarians define as the good life. This image, of course, stands apart from the Faustian concept of progress so dominant in the modern world, "the constant outreaching, the denial of limits, the willingness to dissolve all into endless instrumental activity," to use Richard Weaver's phrase.[34] Not high consumption nor the fastest speed, not the amassing of toys or wealth, but the healthy household stands as the agrarian ideal to which all other goals are subordinate.

When speaking of the good life, however, agrarians tend to avoid generalizations and to turn instead to the particular. The lived agrarian life is full of specific, familiar, tangible things—tools, sheds, barns, tables, fences, gardens, porches, woodlots, shade trees, garages, birdhouses, rock outcroppings, and the like. The people of the household, too, are recognized for their particularities; they are members who belong, for better or worse, not labor and management or producers and consumers. Louis Bromfield was one such agrarian voice, proud of his Ohio farm, Malabar, and pleased to tell people what he had learned there. True significance in life, he came to see, arose not in a romantic retreat to the past but in

> a tangible world of great and insistent reality, made up of such things as houses, and ponds, fertile soils, a beautiful and rich landscape and the friendship and perhaps the respect of my fellow men and fellow farmers.[35]

VII

To complete the picture of the New Agrarianism, one needs finally to address a few misapprehensions about it, to situate it on the contemporary political spectrum, and to take note of its chief opponent.

The New Agrarianism makes no claim that people who till the land are better than others. It would be good if they were, given the importance of their work, and on average they may once have been. But the industrial mind-set has so permeated agriculture today (as Don Kurtz shows here in "The Competitive Edge") that agrarian values are hardly more prevalent on large farms than in suburban neighborhoods.[36] Indeed, to look at the policy platforms of an organization such as the American Farm Bureau Federation is to see, if not a total rejection of agrarian values, at least an attitude toward the land that is industrialized, commodity based, and exploitive. One searches in vain through policy positions for any sense that landowners are responsible for the damage they do or that they are members of land communities with distinct duties as such. The agrarian belief that the market does not cleanse, for instance, is nowhere to be seen. The ideal of the livelihood farm is in no way honored. When Jefferson exalted the virtue of small farmers, he did so precisely because they were independent of the market and sought no special favors from government;[37] today's industrial farmers do not possess these traits.[38]

Nor is the New Agrarianism an attempt to restore a golden age of any sort, for a golden age never existed.[39] Indeed, the coming of the plow to the frontier often brought widespread erosion and waterway degradation when farmers failed to tailor practices to the land's features and limits. On particular farms and in specific neighborhoods, good progress was being made by the early twentieth century in creating a fully realized agrarian culture, one that respected land, promoted economic diversity, and nurtured an ethic of permanence. But that progress was largely undercut by the potent interaction of the market, new technology, and liberal individualism. What New Agrarians exalt today is not a specific era or place but the work of model individual agrarians and the ideals they embody.[40] It is to them and their hopes that one turns today for inspiration.

Also needing clarification is the presumption that agrarianism is inherently anti-science or in all ways anti-market. For agrarians, science is merely a tool, highly useful when well directed (as David Orr shows here) but highly destructive when it is not. What agrarians insist on is that the products of science be tested before implementation to see whether they sustain or undercut sound values and goals. What they condemn are the messianic aura that surrounds science and the flawed assumption that newer and higher technology is always better. As for the market, agrarians view it in analogous ways: As a mechanism for stimulating enterprise and generating overall wealth, it is highly useful—but only when well tamed. Agrarians spurn the fantasy that the market is always or even usually good at measuring values, particularly when it comes to nature. They reject, too, the damaging assumption that a contract-based market can serve well as the dominant connector among people, displacing nonmarket cooperations based on custom, status, faith, and virtue. For agrarians, the critiques of economist Joseph Schumpeter and social theorist Daniel Bell ring true: The capitalist market is a destructive as well as a creative force, and its destruction is most pernicious when it undercuts the very virtues and social relations on which it depends.[41] Agrarians would also agree with historian Eugene Genovese's admonition:

> It is hard to believe that, however much we must accept a market economy, we could expect to live as civilized human beings in a society that makes the market the arbiter of our moral, spiritual, and political life.[42]

One claim not so easily dismissed is that agrarians are neo-Luddites, that is, sympathizers with the fictional Ned (or Edward) Ludd, in whose name craftsmen of the cloth trade rose up in early-nineteenth-century England to challenge the coming of the new industrial order. Agrarians introduced to Luddite ideas might well find them congenial, for the Luddites, too, were concerned about the ravages of market forces on families, communities, leisure, and prevailing moral values. Luddites are remembered today for their occasional symbolic efforts to smash the machines that were disrupting their lives, but the Luddite cause, as historians have explained, was never simply about technology and economic efficiency. It had other, deeper roots; it was a defense of basic values, a defense not of the livelihood farm per se but nonetheless of the household as center of economic productivity. It was a defense of a way of life far better, as many cloth workers saw matters, than the life offered them by the overbearing world of factories.[43]

A final question worth raising is whether the New Agrarianism is best understood in metaphorical terms, not as a constellation of practical proposals, back-to-the-land and otherwise, but as an effort to draw on agrarian traditions to criticize modern culture.[44] Realistically, few American families can return to the land as the primary center of the family economy, although "few" might over time include several million families (as Gene Logsdon contends here in "What Comes Around"). To admit this practical limit, however, is not to say that agrarianism is therefore mere metaphor. With appropriate modifications, agrarian ways are suitable for a far larger number of families. Even those with full-time office jobs can add significant agrarian elements to their lives and locales. Beyond that, the values that form the agrarian core provide a firm perspective from which to interact with the world and pass judgment on its ways. Although agrarian reform is best begun at the individual and family levels, it is easy to imagine that a legislative body guided by agrarian principles would produce far different laws from those now in effect. Agrarian attitudes would bring changes to golf courses, parks, landscape plans, drainage systems, residential developments, production methods, and a wide range of other collective activities—as David Orr shows in his contribution here, "The Urban-Agrarian Mind." Given that so much of the national economy arises from choices in recreation and leisure, a major shift in interest to outdoor, local, nonconsumptive recreation alone could have major effects on the economy and the land.

As for its position on the conservative–liberal political spectrum, the New Agrarianism has no settled home. It listens attentively to those who speak of traditional values, family, and personal responsibility, but it wants little to do with conservative politicians devoted to conserving corporate profits and open markets while caring little about lands, families, and rural communities. The party of Wall Street and global conglomerates can hardly expect agrarian support. As Eugene Genovese has observed, too many conservatives talk of sustaining traditional values but show little recognition of the social relations and community structures necessary for their sustenance. On the liberal side, talk of compassion and inclusiveness strikes responsive chords; in the agrarian household, foibles and weaknesses are no cause for expulsion. Yet in their quest for inclusiveness and institutional reform, liberals too often defend personal immorality and irresponsibility. In the liberal worldview, family and local community often count for little and problems best addressed in the home or neighborhood are readily turned over to institutions and highly paid "experts." Indeed, liberals typically (if unintentionally) reinforce the radical individualism and self-centeredness that agrarians identify as central roots of modern ills.

Among American traditions, the New Agrarianism is most similar to the strand of thought sometimes referred to as Southern conservatism, a tradition ably carried into the post–World War II era by Richard Weaver, distilled and defended recently by Eugene Genovese, and expressed most eloquently today by Wendell Berry. The Southern conservative and agrarian traditions came together in an important essay collection published in 1930, *I'll Take My Stand,* authored by Twelve Southerners.[45] Collectively, the authors expressed alarm over the effects of industrialism and materialism on the mannerly, leisurely, humanistic culture they viewed as the South's greatest treasure. Several of the twelve authors spoke of the land, its physical decline, and the practical prospects it offered for rebuilding the country along lines less grasping and frenetic. Others among them used the livelihood farm less as economic prescription and more as cultural emblem. In the decades since 1930, both the conservative, religious humanism of *I'll Take My Stand* and the book's agrarian ideals have taken a severe beating. Yet the heartbeat has remained. And as the writings gathered here attest, they are slowly regaining strength.

I'll Take My Stand deserves mention for the additional reason that, in it and in the making of it, the authors struggled to capture in a word or phrase the opposing culture that bore down on them with such force, the culture that since the early nineteenth century had relentlessly sapped strength from older agrarian forms. Drawing on the tradition dating back to Thomas Carlyle, they ultimately chose the term *industrialism* ("the fiery gnawing of industrialism," as one of the twelve, Donald Davidson, had earlier expressed it),[46] by which they meant a full constellation of values and methods of social organization rather than a particular means of accumulating capital and producing physical goods. As John Crowe Ransom put it in his contribution to *I'll Take My Stand,* industrialism was "the latest form of pioneering and the worst," its driving energy the "principle of boundless aggression against nature." Although he admitted that the industrial mind displayed "almost miraculous cunning," it was, he urged, "rightly a menial": "It needs to be strongly governed or it will destroy the economy of the household."[47]

Several of the authors of *I'll Take My Stand* preferred to characterize the opposing force as communism, a term they also vested with special meaning.[48] The dominant, aggressive culture, they believed, was fragmenting the social and moral order and creating the atomistic mass man; in the process, it was overpowering local arrangements and diversities and reducing humankind to a base, common existence. In their individual essays, the authors for the most part employed more poetic expressions: The enemy was the "Kingdom of Whirl," as Henry Kline put it, "the culture of aimless flux."[49] Andrew Lytle, the truest agrarian among the twelve, used even sharper words of condemnation: "We have been slobbered upon by those who have chewed the mad root's poison, a poison which permeates to the spirit and rots the soul."[50]

In later writings, Allen Tate and John Crowe Ransom often used the term *capitalism* rather than *industrialism*. After World War II, Richard Weaver crafted and used the expression "universal materialism and technification."[51] In the writings presented here, Wendell Berry and David Orr also use the term *industrialism,* employing it in the same suggestive, encompassing way as did the Twelve Southerners of 1930. Historian Donald Worster, in his analysis of the cultural roots of land degradation, finds the term *materialism* more apt in capturing this powerful, complex phenomenon. Describing his work in urban communities, Alan Durning draws on the terms *possessiveness* and *consumption* to label the corrosive elements within us. In all cases, though, the contrasting worldviews are similarly set forth: mastery over nature versus harmonious living within it; nature as collection of resources versus nature as organic whole; place as incidental versus place as essential; knowledge as sufficient versus knowledge as radically incomplete; value largely in exchange versus value largely in use; unlimited wants versus manageable needs; a throughput, throwaway mentality versus a cyclical, fertility-restoring mentality; labor as means to consumption versus labor as integral to the good life; morals as largely self-selected versus morals as cultural inheritance; property as commodity versus property as propriety; the future as heavily discounted versus the future as weighty consideration; communities as mere convenience versus communities as cultural essentials; the household as place of consumption versus the household as center of life; life as fragmented versus life as a unified whole.

VIII

Writing late in his life, Aldo Leopold bemoaned the reality that the average modern of his day (the 1940s) had "lost his rootage in the land." The "shallow-minded modern," he penned,

assumes that he has already discovered what is important; it is such who prate of empires, political or economic, that will last a thousand years. It is only the
scholar who appreciates that all history consists of
successive excursions from a single starting-point, to
which man returns again and again to organize yet
another search for a durable scale of values.[52]

Pondering the challenges of promoting healthy landscapes, as
he did for many years, Leopold came to see that the core problem
lay within the human heart and soul. People simply did not perceive the land and recognize their ties to it; they failed to love the
land as they ought; they failed to understand that their dealings
with the land were, at bottom, not matters of expediency alone but
of ethics as well. It was a sobering conclusion that Leopold reached,
for the transformation of ethics, he knew, was hardly the work of a
single lifetime. And, as he knew, "no important change in ethics
was ever accomplished without an internal change in our intellectual emphasis, loyalties, affections, and convictions."[53]

On every page, the essays here gathered reflect these yearnings,
and in them one approaches the heart and soul of the New Agrarianism: yearnings to regain society's rootage in the land; yearnings
to stimulate sounder loyalties, affections, and convictions; yearnings, in the end, to craft a scale of values more likely to endure.

PART I

NEW PROSPECTS

Chapter 1

Learning from the Prairie

Scott Russell Sanders

Many of the New Agrarianism's concerns center on the land community, its long-term health, and how people live within it. Too many human ways of drawing food, fiber, and minerals from the land are, in the long run, destructive of land and people. New production methods are needed, ways that are more sensitive to nature's limits and more respectful of its processes and mysteries.

Scattered around the country—chiefly outside major universities and industrial research centers—agrarian-minded scientists are at work studying nature and searching for ways to draw sustenance from the land without eroding its soils, polluting its waters, destroying its wildlife, and diminishing its natural beauty. A particularly promising aspect of that work goes on at The Land Institute, near Salina, Kansas, where Dr. Wes Jackson heads a team of researchers studying the traits of the native tall-grass prairie to gain insights on developing sustainable farming methods tailored to local conditions. In the work at The Land Institute one sees the intelligence, vision, love of nature, and dedication to place that characterize agrarian enterprises everywhere.

This portrait of The Land Institute and its guiding leaders was written by award-winning author Scott Russell Sanders, many of whose works embrace an agrarian perspective.

In Salina, Kansas, first thing in the morning on the last day of October, not much is stirring except pickup trucks and rain. Pumpkins balanced on porch railings gleam in the streetlights. Scarecrows and skeletons loom in yards out front of low frame houses. Tonight

3

the children of Salina will troop from door to door in costumes, begging candy. But this morning, only a few of their grandparents cruise the wet streets in search of breakfast.

In the diner where I come to rest, the average age of the customers is around seventy, and the talk is mainly about family, politics, and prices. Beef sells for less than the cost of raising it. There's a glut of soybeans and wheat. More local farmers have fallen sick from handling those blasted chemicals. More have gone bankrupt.

When a waitress in a leopard suit arrives to take an order from the booth next to mine, a portly man greets her by complaining that Halloween has turned out wet. "It's a true upset to me," the man says. "Last year I had two hundred children ring my bell." The waitress calls him honey and sympathizes.

An older woman bustles in from the street, tugs a scarf from her helmet of white curls, and declares to everyone in the diner, "Who says it can't rain in Kansas?"

At the counter, a woman wearing a sweatshirt emblazoned with three bears swivels around on her stool. "Oh, it rains every once in a while," she replies, "and when it does, look out!"

Here in the heart of Kansas, where tallgrass prairie gives way to midgrass, about twenty-nine inches of water fall every year, enough to keep the pastures thick and lure farmers into planting row-crops. Like farmers elsewhere, they spray pesticides and herbicides, spread artificial fertilizer, and irrigate in dry weather. They plow and plant and harvest using heavy machinery that runs on petroleum. They do everything the land-grant colleges and agribusinesses tell them to do, and still many of them go broke. And every year, from every plowed acre in Kansas, an average of two to eight tons of topsoil wash away. The streams near Salina carry rich dirt and troubling chemicals into the Missouri River, then to the Mississippi, and eventually to the Gulf of Mexico.

Industrial agriculture puts food on our tables and on the tables of much of the rest of the world. But the land and farmers pay a terrible price, and so do all the species that depend on the land, including us.

I've come to Salina to speak with a man who's seeking a radical remedy for all of that—literally radical, one that goes back to the roots, of plants and of agriculture. Over the past six or eight years I've bumped into Wes Jackson several times at gatherings of folks who worry about the earth's future, but this is my first visit to his home ground. Wes has been here since 1976, when he and his then-wife, Dana, founded The Land Institute, a place devoted to finding out how we can provide food, shelter, and energy without degrading the planet. He won a MacArthur fellowship in 1992 for his efforts, and he has begun to win support in the scientific community for a revolutionary approach to farming that he calls perennial polyculture—crops intermingled in a field that is never plowed, because the plants grow back on their own every year. The goal of this grand experiment is to create a form of agriculture that, like a prairie, runs entirely on sunlight and rain.

❊   ❊   ❊

To reach The Land Institute, I drive past grain silos lined up in rows like the columns of a great cathedral; they are lit this early morning by security lights, their tops barely distinguishable from the murky sky. I drive past warehouses, truck stops, motels, fast-food emporiums, lots full of RVs and modular homes; past a clump of sunflowers blooming in a fence-corner at the turn-off for Wal-Mart; past filling stations where gas sells for eighty-five cents a gallon. The windshield wipers can't keep up with the rain.

When pavement gives way to gravel, I pass a feedlot where a hundred or so cattle stand in mud and lap grain from troughs. Since entering Kansas, I've seen billboards urging everyone to eat more beef, but the sight of these animals wallowing in a churned-up rectangle of mud does not stimulate my appetite. The feedlot is enclosed by electrified wire strung on crooked fence posts made from Osage orange trees. In a hedgerow nearby, living Osage oranges have begun to drop their yellow fruits, which are the size of grapefruits but with a bumpy surface like that of the human brain. After the road crosses the Smoky Hill River, it leaves the flat bottomland, where bright green shoots of alfalfa and winter wheat sprout from dirt the color of chocolate, then climbs up onto a rolling prairie, where the Land Institute occupies 370 acres.

Wes Jackson meets me in the yellow brick house that serves for an office. It's easy to believe he played football at Kansas Wesleyan, because he's a burly man, with a broad, outdoor face leathered by sun and a full head of steel-gray hair. Although he'll soon be able to collect Social Security, he looks a decade younger. He wears a flannel shirt the shade of mulberries, blue jeans, and black leather boots that have quite a few miles on them. For a man who thinks we've been farming the wrong way for about 10,000 years, he laughs often and delights in much. He also talks readily and well, with a prairie drawl acquired while growing up on a farm in the Kansas River Valley, over near Topeka.

"I'm glad you found your way all right," he says. "Can't hide a thing out here on the prairie, but you'd be surprised at the people who get lost."

When I admit to having asked directions at a station that advertised gas for eighty-five cents a gallon, he tells me, "The price of gasoline is a symptom of our capacity for denial. We pay for gas based on how much of it is above ground, not how much is left below. We ignore its real scarcity."

Wes and I sit at the kitchen table while coffee perks, a copy machine on one side of us, a wood stove on the other. The walls are lined with shelves bearing jars full of seeds. Every now and again I ask a question, but mainly I listen. Wes talks in a voice as big as he is, all the while fixing me with a steady gaze through wire-rimmed spectacles, to make sure I'm following.

He points out that our whole economy rides on cheap oil, which he calls "fossil sunlight," and nowhere is this dependence more evident than in agriculture. Natural gas is the raw material for anhydrous ammonia, which farmers spread on fields to compensate for the loss of natural fertility. We hammer the soil, he says, then put it on life support. We replace draft horses and hand labor with diesel-powered machines. We replace the small-scale farming of mixed crops with vast plantations of single crops, usually hybrids, which are so poorly adapted that we have to protect them from weeds and pests with heavy doses of petroleum-based poisons.

While cheap oil has accelerated our journey down the wrong path, we set out on that path long before we discovered the convenience of fossil sunlight, according to Wes. Our ancestors made the key mistake at the very beginnings of agriculture, when they started digging up the fields and baring the soil. The great river civilizations along the Tigris, Euphrates, Ganges, and Nile could get away with that for a while, since floods kept bringing in fresh dirt. But as populations expanded and tillage crept out of the river bottoms into the hills, the soil began to wash away.

"The neolithic farmers began mining ecological capital," he explains. "That was the true Fall, worse than anything poor Eve might have done."

Wes knows his Bible, and he draws from history and philosophy and literature as easily as from plant genetics, the field in which he earned his Ph.D. at North Carolina State. At one point he quotes a famous phrase from the prophet Isaiah, then questions whether we're actually better off beating swords into plowshares. Wes is wary of swords, but also wary of plows. Where our ancestors went wrong, he believes, was in choosing to cultivate annual crops, which have to be planted each year in newly turned soil. The choice is understandable, since annual plants take hold more quickly and bear more abundantly than perennials do, and our ancestors had no way of measuring the long-term consequences of all that digging and tilling.

But what's the alternative? How else can we feed ourselves?

Wes takes me outside to look at the radically different model for agriculture that he's been studying for more than twenty years—the native prairie. Because the rain hasn't let up, we drive a short distance along the road in his battered Toyota pickup, then pass through a gate and go jouncing onto an eighty-acre stretch of prairie that's never been plowed. The rusty, swaying stalks of big bluestem wave higher than the windshield. The shorter stalks of little bluestem, Indian grass, and switchgrass brush against the fenders. We stop on the highest ridge and roll down the windows so rain blows on our faces, and we gaze across a rippling, sensuous landscape, all rounded flanks and shadowy crevices.

"This would be a fine spot for the Second Coming," Wes murmurs. After a pause he adds, "Not that we need saving here in Kansas."

The grasses are like a luxurious covering of fur, tinted copper and silver and gold. In spring or summer this place would be fiercely green and spangled with flowers, vibrant with butterflies and songbirds. Now, in the fall, Wes reports, it's thick with pheasant, quail, and wild turkey. He and his colleagues don't harvest seeds here, but they do burn the prairie once every two or three years, and they keep it grazed with Texas longhorns, whose bellows we can hear now and again over the purr of engine and rain. Eventually the cattle will give way to bison, a species better adapted to these grasslands. From the pickup, we can see a few bison browsing on a neighbor's land, their shaggy coats dark with rain.

In every season, the prairie is lovely beyond words. It supports a wealth of wildlife, resists diseases and pests, holds water, recycles fibers, fixes nitrogen, builds soil. And it achieves all of that while using only sunlight, air, snow, and rain. If we hope to achieve as much in our agriculture, Wes argues, then we'd better study how the prairie works. Not just the Kansas prairie, but every one we know about elsewhere, works by combining four basic types of perennial plants—warm-season grasses, cool-season grasses, legumes, and sunflowers—all growing back year after year from the roots. The soil is never laid bare. The prairie survives droughts and floods and insects and pathogens because the long winnowing process of evolution has adapted the plant communities to local conditions.

"The earth is an ecological mosaic," Wes explains. "We're only beginning to recognize the powers inherent in local adaptation."

If you wish to draw on that natural wisdom in agriculture, he tells me as we drive toward the greenhouse, then here in Kansas you need to mimic the structure of the prairie. It's all the more crucial a model, he figures, because at least 70 percent of the calories that humans eat come directly or indirectly from grains, and all our grains started as wild grasses. For nearly a quarter-century, Wes and his colleagues have been working to develop what he calls perennial polyculture—as opposed to the annual monoculture of traditional farming—by experimenting with mixtures of wild plants. Recently they've focused on Illinois bundleflower, a nitrogen-fixing legume whose seed is about 38 percent protein; *Leymus,* a mammoth wild rye; eastern gama grass, a bunchgrass that's related to corn but is three times as rich in protein; and Maximilian sunflower, a plentiful source of oil.

In the sweet-smelling greenhouse, we find seeds from these and other plants drying in paper bags clipped to lines with clothes pins. The bags are marked so as to identify the plots outside where the seeds were gathered; each plot represents a distinct ecological community. Over the years, researchers at the Land Institute have experimented with hundreds of combinations, seeking to answer four fundamental questions, which Wes recites for me in a near-shout as rain hammers down on the greenhouse roof: Can perennial grains, which invest so much in roots, also produce high yields of seed? Can perennial species yield more when planted in combination with other species, as on the prairie, than when planted alone? Can a perennial polyculture meet its own needs for nitrogen? Can it adequately manage weeds and insects and disease?

So far, Wes believes, they can answer a tentative yes to all those questions. For example, his daughter Laura, now a professor of biology at the University of Northern Iowa, has identified a mutant strain of eastern gama grass whose seed production is four times greater than normal—without any corresponding loss of root mass or vigor.

More and more scientists are now testing this approach. After returning home from Salina, I'll contact Stephen Jones at Washington State University, a plant geneticist who is developing perennial forms of wheat suited to the dry soils of his region. I'll correspond with a colleague of Jones's at Washington State, John Reganold, a professor of soil science who predicts that with these design-by-nature methods, "soil quality will significantly improve—better structure, more organic matter, increased biological activity, and thicker topsoil." I'll learn about efforts in the Philippines to develop perennial forms of rice. I'll speak with the director of the plant-biotechnology program at the University of Georgia, Andrew Paterson, who is also experimenting with perennial grains. I'll contact Stuart Pimm at the University of Tennessee, a conservation biologist who has reported in the journal *Nature* on Land Institute experiments that show that mixtures of wild plants not only rival monocultures in productivity but also inhibit weeds and resist pathogens while building fertility.

I'll contact all those people, and more, after returning home from Salina. But right now I'm listening to the fervent voice of Wes Jackson, who's lamenting that the United States loses 2 billion tons of topsoil a year to erosion. The cost of that—in pollution of waterways, silting of reservoirs, and lost productivity—is $40 billion, according to the U.S. Department of Agriculture. Wes estimates that only 50 million of the 400 million tillable acres in the United States are flatland, and even those are susceptible to erosion. The remaining 350 million acres—seven-eighths of the total—range from mildly to highly erodible, and thus are prime territory for perennial polyculture.

He flings these statistics at me as we drive into Salina for lunch at a Mexican restaurant. Maybe what set him hungering for Mexican food were the strings of bright red jalapeño peppers hanging in the greenhouse among the brown paper sacks full of seeds. Whatever the inspiration, Wes launches into his plateful of burritos with the zeal of a man who has done a hard morning's work. As we eat, a nearby television broadcasts a game between Kansas State and the University of Kansas. Checking the score, Wes explains, "My nephew plays for KU at guard, my old position." When he learns that Kansas is losing, he turns his back to the TV and resumes telling me about what he calls natural-systems agriculture.

"The old paradigm," he says, "is the industrial model, which figures we can beat nature, make it dance to our tune, use up whatever we need and dump our wastes wherever's convenient. The new paradigm, the one we're following at the Land, believes less in human cleverness and more in natural wisdom. The prairie knows what it's doing—it's been trying things out for a long while—and so we've made ourselves students of the prairie."

Transforming perennial polyculture from a research program into a feasible alternative for the working farmer will require many more years of painstaking effort, Wes admits. Researchers must breed high-yielding varieties of perennial grains and discover combinations of species that rival the productivity of the wild prairie. Engineers must design machinery for harvesting mixed grains that may ripen at different times. Farmers must be persuaded to try the new seeds and new practices, and consumers must be persuaded to eat unfamiliar foods.

In keeping with his mission, before we leave the Mexican restaurant Wes urges me to try the whole wheat tortilla chips. "They're a lot tastier than the cornmeal, don't you think?"

I try them, and I agree.

It's still raining when we climb back into the pickup, and as we drive into the countryside Wes keeps shaking his head at the black slurry pouring off the fields. "That's gold running away," he says.

"Farmers are always worrying about money, and right there's pure wealth just washing away. It takes up to a thousand years to make an inch of topsoil."

He goes on to speak about the need for training farmers, a subject close to his heart. "The children in rural schools are one day going to be in charge of the 400 million acres of tillable land in this country. So they'll have the greatest ecological impact of any group." To help inform those schools—and help resettle the small towns in which many of those children will grow up—the Land Institute has created a Rural Community Studies Center in Matfield Green, a tiny settlement in the Flint Hills about a hundred miles southeast of Salina. "We want to bring the message of ecology to bear on the curriculum of rural schools," he says. "I want those young people to go to Kansas State, Ohio State, all the ag schools, and ask questions that push beyond the existing paradigm."

How well would annual monoculture perform if it weren't subsidized by inputs of petroleum and groundwater, and if it weren't allowed to write off the ecological costs of pesticides and herbicides and erosion? To answer that question, the Land Institute has devoted 150 acres to the Sunshine Farm, a ten-year project for growing livestock and conventional crops without fossil fuels, chemicals, or irrigation. The Sunshine Farm is where we go next, and the arrival of our truck wakes three dappled-gray Percheron draft horses from their rainy drowse in a paddock beside the barn. For the heaviest work there's also a tractor, but it shelters inside the barn and it runs on bio-diesel fuel made from soybeans and sunflower seeds. The farmhouse is heated with wood, and all the buildings are lit from batteries charged by a bank of photovoltaic cells.

Six years into the study, data from the Sunshine Farm are providing a truer measure of how much conventional farming costs. Marty Bender, who manages the farm, explains, "We look at the energy content of all the crops and livestock that we produce, and we look at the inputs—fuels, feeds, stock, seeds, tools, labor. If you divide our outputs by our inputs, the ratio is comparable to what you see on Amish farms. And that tells me we're on the right track."

"When all the numbers are in," Wes predicts, "I'm sure the prairie's way will beat the pants off the industrial way."

✤   ✤   ✤

Back in the yellow brick office, Wes unrolls onto a table what he calls the Big Chart, which lays out a twenty-five-year research plan. The boxes on the chart frame problems to be solved, and the arrows all point toward the vision of a sustainable agriculture that will overturn the mistaken practices of the past ten millennia. It's a bold scheme. Already, scientists like Stephen Jones, Andrew Paterson, John Reganold, and Laura Jackson have begun to work on pieces of the puzzle. With half a dozen full-time investigators and their assistants, plus eight student interns and five or six graduate students each year, the Land Institute operates now with an annual budget of $850,000, supported by foundations and private donors and the tireless labor of many friends.

This endeavor, now almost a quarter-century old, nearly died in infancy. As a young man with a family, Wes gave up a tenured position at California State in order to homestead in Kansas, then put every cent he had into starting the Land Institute. "Six months later," he recalls, "our only building burnt down, with all our books and tools. A great darkness came over me. It seemed like the world was telling me to quit. But if you're raised on a farm you're used to making things work. If you don't get it right the first time, you have another go at it. So we rebuilt."

To carry on the necessary future research, Wes calculates they'll need between $5 million and $7 million a year—not much money when you consider that estimated yearly loss of $40 billion from soil erosion in the United States. This higher lever of funding can only come with backing from the U.S. Department of Agriculture and even from agribusiness firms. "So far," he admits, "we've hit a brick wall at USDA. When you talk with them about learning from the prairie, following nature as measure and pattern, their eyes glaze over."

He realizes how difficult it will be to pry money from institutions whose philosophy of farming he so squarely opposes, but he relishes the challenge. "In America," he tells me as I prepare to leave, "we've got mostly two kinds of scientists—the ones who get us in trouble, and the ones who tell us what the troubles are—but very few who are looking for solutions. Here at the Land Institute, we're looking for solutions."

Before I go, I can't help asking him to explain how a Kansas farm boy grew up to become a visionary who's trying to revolutionize farming. He can't say for sure. His family's been in Kansas since 1854 (the year that *Walden* was published). His great-grandfather fought alongside John Brown at the Battle of Blackjack Creek, against proslavery hooligans from Missouri. His grandchildren are the sixth generation to live in the state. So he feels committed to this region for the long haul, and he wants it to be a beautiful and fertile place well after he's gone. "It seems like, no matter what else I tried, I just kept thinking about the source—soil, water, photosynthesis, the things that sustain us." Is he hopeful that a durable form of agriculture will be found in time to feed the earth's swelling population? "We don't know how this is all going to turn out," he admits. "But the risky thing is to do nothing, to keep on going the way we've been going. No matter how dark the times, it's still worthwhile to do good work."

❧   ❧   ❧

The next morning, as I drive east through even heavier rain toward my home in Indiana, the radio carries reports of brimming rivers and flooded roads across Kansas. The plowed fields I pass are gouged by rivulets and the roadside ditches run black with dirt. But where grass covers the land, there's no sign of runoff, for the prairie keeps doing what it's learned how to do over thousands of years—holding water, building soil, waiting for spring.

Chapter 2

Linking Tables to Farms

Dan Imhoff

A particularly visible strand of the New Agrarianism is the movement known as community-supported agriculture, or CSA, which since its North American introduction in 1986 has spread rapidly throughout the United States and Canada. In a typical CSA arrangement, a gathering of nonfarm families and individuals contracts with a farmer to produce vegetables, fruits, and other foods. Participating members share the costs of the arrangement. They also share the food that is produced, which means they shoulder collectively many of the risks—of weather, pests, and the like—that make farming such a challenging life. In many projects, CSA members contribute labor as well as cash to the enterprise, working in fields, gathering and sorting produce, and arranging for deliveries and pickups. Almost always in these arrangements, farming methods use few or no inorganic fertilizers and pesticides. By guaranteeing a market, CSA projects provide farmers with reliable income and reduce attendant risks. They also provide an outlet for produce that is blemished or otherwise imperfect in appearance.

In the following essay, apple-grower and writer Dan Imhoff explains why he and his family participated in a CSA project in their home region of northern California. He explores the benefits and challenges of such arrangements, not just for farmers and participating city dwellers but also for the land itself. As Imhoff observes in his final sentence, CSA projects bring urban eaters closer to the soil than they've been in a long time.

Until our move to a rural valley, my family's primary provider of fruits and vegetables for four years was a northern California farmer named Kathy Barsotti. Every Tuesday a half-bushel box arrived on our doorstep, fresh from the fields and orchards at Capay Fruits and Vegetables, eighty miles northeast of San Francisco. While my wife or I sorted through the delivery, the other read aloud from the accompanying one-page newsletter, which detailed the contents, related weekly weather and farm progress, and offered intriguing recipe ideas. One week in early April, Barsotti wrote:

> The tomatoes that we transplanted last week look very nice, thank goodness! It was a real blow to lose our first planting. We have fruit on the Asian pears, the figs, the apricots, and the peaches. As soon as we get a chance, we need to begin thinning so we get a decent size on the ripe ones. All of our thinning is done by hand. The mandarins are putting out their first flush of leaves, and the flowers are just starting to bloom. Soon we will be living in the heavenly fragrance of orange blossoms.

Twenty dollars per week (plus a five-dollar delivery fee split among neighbors) supplied us and our two small children with most of the produce we needed, including leftovers for a weekly soup. Our son's first solid food came from these boxes: steamed vegetables and fresh fruit that we mashed with a small hand-mill. In winter, leafy greens, root crops, and citrus fruit predominated. By midsummer, the boxes overflowed. We had relinquished our choice in foods, learning instead to make the most of what Kathy managed to produce herself or procure from neighboring farmers during any given week. In doing so, we were participating in community-supported agriculture (CSA), a movement that since the mid-1980s has aided many small, diverse farms across the country in their struggles with rising land values, expanding industrial organic farms, government aid for agribusiness, and increasing production costs.

According to Trauger Groh and Steven McFadden, authors of *Farms of Tomorrow Revisited*,[1] the United States in 1997 had upward of 1,000 farms that, like Capay Fruits and Vegetables, offered a direct relationship with a body of members. Individually, each of these farms is referred to as a CSA. Although common goals unite them, CSAs vary in many ways. Some are best described as "mutual farms," in which members enter into a partnership with the farmer, investing $450–$900 for shares at the beginning of the year in return for edible dividends throughout the growing season. Others are monthly subscription arrangements with an emphasis on customer service, such as provision of prewashed salad mix, fresh-cut flowers, and home delivery. CSAs range from 5-share gardens cultivated solely by members to large farms serving nearly 1,000 shareholders. Although the majority rely on designated drop-off sites where customers retrieve their weekly bounty, some insist that pick-your-own programs are what separate pure CSA farms from mere subscription arrangements and other less ideologically driven interpretations. In terms of a national average, a weekly CSA share provides enough food for three people and costs between $12 and $25.

"The CSA is a valuable model for many reasons," explained Judith Redmond, executive director of the Community Alliance with Family Farmers (CAFF) in Davis, California, and cofounder of Full Belly Farms, which has operated a CSA in the Capay Valley since 1994. "Because most members pay a yearly share up front before the growing season begins, it significantly helps out with our cash flow. And because our members expect to receive what's done well on the farm on any given week, we have a guaranteed outlet for in-season produce and heavy yielders."

For my wife and me, CSA membership was a lifestyle choice, not much different from bicycling to work, using a cotton diaper service, or composting in a backyard worm box. The challenge of using only what was in season, combined with the fresh, organically grown ingredients, elevated our culinary repertoires. We became creative with kale and kohlrabi, looked forward to weeks of leeks, learned to live without tomatoes and green beans in winter and spring, and evolved into competent soup makers.

I visited Kathy Barsotti's farm on a sizzling May afternoon. The Capay Valley has a broad, flat belly that spreads gracefully and is surrounded by hills with grassy meadows and oak groves. The farm was as pastoral a scene as one could imagine, though so much pollen was in the air that my head pounded. A small, fit woman with a welcoming smile, Barsotti was watering tomato starts in the greenhouse when I arrived. A safari helmet shielded her face from the sun. Her three full-time farmhands, who I'd been reading about for some time, were thinning peaches.

Barsotti owned twenty-five acres and farmed a total of ninety. The CSA resembled an extensive orchard and garden and was producing twenty different vegetables. Each row hosted one or more crops: ambrosia melons and Mickey Lee watermelons; Green Zebra and Thessaloniki tomatoes; romaine and red leaf lettuce. "This is more like quilting than farming," said Barsotti. "This is what I wanted when I started farming twenty years ago, and running a CSA now allows me to earn my living off ninety acres."

The Capay Valley has in fact been a haven for small, diverse organic farming operations such as Barsotti's. Four other CSAs in the area deliver boxes to San Francisco Bay Area residents in addition to selling wholesale, in farmers' markets, and to restaurants. Although these local farmers remain competitive on some level, a great deal of cooperation among them takes place. "Having two or three CSAs in a region gives a real boost to a community's economy," explained Barsotti, who had been bartering and buying apricots, oranges, salad greens, and a variety of other items from neighboring farmers to ensure that her weekly delivery was as fair to her members as possible.

Fellow valley farmer Judith Redmond concurred. Twice in the past four years, Full Belly had shared its waiting list with other CSAs needing to grow a body of customers. (A great majority of the CSA founders I talked to had waiting lists throughout the 1990s.) I contrasted for a moment that generous gesture with the proprietary nature of agribusiness—in particular, the great rush to patent nearly all the biological wealth on the planet and thereby corner the market on it—and caught a glimmer of the spirit that has cloaked the organic farming movement since its modern prac-titioners in the 1970s decided idealistically to take charge of their food.

Ironically, on my way out of the Capay Valley, I found few opportunities to eat or buy the outstanding food being grown in this productive region—not in the corner grocery store or at the local restaurant. Farming remains a market-oriented system almost everywhere. Crops are raised primarily for export while food is imported for consumption and for centralized processing. I've often read (and even cited) the 1983 statistic that the average food item in the United States travels 1,300 miles before it reaches the dinner table.[2] (Taking into account the North American Free Trade Agreement and the increasing reliance on Chilean imports, that estimate could be conservative today.) CSAs are working to diminish that distance. Most memberships are located within 100 miles of a farm. By limiting transportation and perishability—two factors that constrain what crops can be grown for distant mar-kets—CSAs are able to offer heirloom varieties that taste better, stay fresher longer, and don't have to be fumigated, irradiated, or refrigerated.

Distinctions between locally centered and national or interna-tional farming systems were highlighted in 1997 when two Capay Valley farmers experienced run-ins with local bureaucracies. Both farms, operating in different communities, employed the common practice of using a member's porch as a weekly drop-off site. Aggrieved neighbors, angry with the tenants responsible for creat-ing the drop-off sites, filed complaints. In one instance, a city health department objected to the use of a porch as a delivery location, not citing a specific code provision but insinuating that uninspected produce might raise awkward sanitation concerns. In the other, a business zoning violation was noted, with the implication that local tax or license laws might apply. Both cases left CSA owners feeling vulnerable, fearful that their operations—even though they involved freshly picked, carefully washed, and appropriately packed produce—might one day be curtailed by inflexible health and zoning bureaucracies.

"The lesson we learned," said one of the farmers, "was that we had to take a proactive and friendly approach to any community we were going into. That means not blocking traffic with our delivery trucks—even though couriers do it regularly—and giving out free boxes to neighbors to educate them about community-supported agriculture."

Although a middleman-free, direct-delivery system does forge new links between growers and consumers, CSA leaders often believe that member involvement needs to extend beyond monetary participation. Many CSAs operate and are able to thrive because of the efforts of a core group of members who actively assist with the organizational tasks needed to keep the trains of these detailed operations running on time: bookkeeping, delivery, harvesting, packaging, and communication. More common on the East Coast and in the Midwest, the core-group strategy highlights the fact that CSAs entail a great deal of work besides growing and harvesting. Tedious accounting systems, newsletter writing, weekly planting schedules—all add greatly to the workload of harried growers. By taking over many tasks, a core group can be successful in letting the farmer farm. Help also comes from apprentices, who are a common feature of most farms. Other farmers develop computerized bookkeeping systems, establish toll-free telephone numbers, and cultivate large memberships to keep operations manageable and profitable.

Many CSAs move beyond volunteer arrangements to insist that members shoulder a proportion of the labor. One example is the Genesee Valley CSA outside Rochester, New York, which not only has an active core group of 21 members but also requires that each of its 200 shareholders spends twelve hours per season helping on the farm and four hours assisting with distribution. This kind of hands-on participation differs from the semiannual workdays and harvest festivals that most CSAs sponsor, events that require considerable organization and supervision by farmers and probably result in less work getting done, however beneficial they are in other ways. At such an event at Live Power Community Farm in Covelo, California, I plowed a field behind a pair of Belgian draft horses with biodynamic farmer Steven Decatur at the reins. I struggled to keep the harrow straight as the rich, dark earth unfurled behind me and a flock of blackbirds trailed us, pecking at the newly exposed worms. We worked hard until dusk, the experience seeping into tired muscles. Although Decatur no doubt could have done the work in less time and more effectively, it was an experience that changed me, deepening my appreciation for those people whose passion is farming.

In 1996, core-group members at the Live Power Farm tackled head-on one of the most difficult challenges facing CSA founders across the country: land tenure. Without ownership, farmers can spend a great deal of their income making rent or mortgage payments for lands that carry high prices because of their suitability for supermarkets or housing developments. Live Power's core group spearheaded a campaign that raised nearly $100,000 to purchase the land being used by biodynamic farmers Steve and Gloria Decatur. Within a year, Live Power Farm was placed into a conservation easement, requiring that future uses of the forty-acre property remain agricultural. The Decaturs received agricultural rights to the farm—not its real estate development value. The separation of agricultural and development values gave this hardworking farm family equity in the land they were dedicated to improving. It also set up a procedure whereby the Decaturs can sell their farming title to a trust when they retire or pass it on to their children if they agree to farm it.

"The CSA model is attracting new sources of capital to the issues of small, diverse agriculture," explained Chuck Matthei of Equity Trust, Inc. in Voluntown, Connecticut. Matthei worked on the Live Power conservation arrangement, and he continues to create new legal and technical tools to preserve small farms in public trusts. His "Gaining Ground" seminars, conducted around the country, are aimed at attracting new players to the challenges of vanishing urban cropland, farm inheritance, and conservation. "Because of the CSA model, there is an openness now in public circles to the idea that small farms must be protected," he said. "While its overall percentage of the current food system remains small, community-supported agriculture is having an effect far beyond its size."

It is easy to wax optimistic and idealistic about the CSA model, and wild claims have appeared, citing revolutionary overhauls of the small-farm scene. CSAs continue to cater mostly to well-to-do city dwellers rather than rural residents; the same people who can afford to buy microbrews can now purchase handcrafted fruits and vegetables and feel good about them. Viewed even more skeptically, these farms-in-a-box schemes could be seen as just another form of entertainment, in this case for people who have the time and tools to prepare high-quality meals. Critics also argue that low-income families and farmworkers are shut out of the movement because of the hardship of paying cash at the beginning of a growing season—or even in any given week—and that these people as much as others are in need of fresh, healthful food. Others point to CSA arrangements in which members have little connection to the work at all, other than writing a check and reading weekly newsletters. Many farmers still live on low incomes, and relationships among farming couples are sometimes as strained as ever. Member attrition can be problematic. And even successful farms struggle with undercapitalization.

Across the broad landscape of CSA organizations, however, there seems to be a farm somewhere experimenting with a mechanism to address each of these challenges and criticisms. Indeed, in the case of some CSAs, social and economic missions are as important as food production. Full Belly Farm offers members a chance to sponsor boxes that are donated to single mothers. Sixth-grade teachers at Martin Luther King Middle School in Berkeley, California, use a weekly box to teach students about farming and nutrition. (The children in turn go home and influence their parents, who reflect diverse ethnic and economic backgrounds.) The Rural Development Center in Salinas, California, runs a CSA solely for Latino farmworkers to improve the quality of foods available to them. Steve Smith, a third-generation farmer in Bedford, Kentucky, believes the CSA model might help keep tobacco farmers in his region from having to sell their family farms. The members of yet another CSA farm congregate annually one day before the growing season to hear the budget presented and then to decide communally how much each will contribute, according to ability to pay.

The Food Bank Farm, on the confluence of the Connecticut and Fort Rivers in Hadley, Massachusetts, donates half of its weekly output to needy families, soup kitchens, halfway houses, and other organizations in the area. Even while funding these charitable outreach efforts, members receive culinary dividends that are reputedly a bargain. According to one survey, the Food Bank Farm's yearly produce share, priced at $375, would have cost $800 if bought piecemeal at Stop & Shop, the regional conventional supermarket, or $1,200 if bought at Bread & Circus, an organic foods market.

The CSA movement began in the United States in 1986 when Indian Line Farm first offered shares in its vegetable harvest in Great Barrington, Massachusetts. Indian Line's founder, the late Robyn Van En, learned the idea from Jan Van der Tuin, an American farmer credited with coining the term *community-supported agriculture*. Van der Tuin in turn had learned of it while living in Switzerland. According to historical accounts, farmer–consumer arrangements of the CSA type existed in various parts of Europe in the 1970s and in Japan in the decade before. Since crossing the ocean, CSAs have sprouted up almost exponentially in the United

States and in Canada, where at least sixty-three were reported in 1996. They range from tiny 5-member backyard gardens in San Francisco's Mission District to 800-share operations such as Angelic Organics outside Chicago and Be Wise Ranch in San Diego. Today, CSAs reach as far north as Anchorage, Alaska, span the breadth of Canada, and sweep diagonally across the continent and down into Florida.

With a conservative back-of-the-napkin calculation, it is probably fair to estimate that CSAs account for 1 percent or a bit more of sales of organically grown produce in the United States, or $30–$40 million. Although CSAs are a mere brush stroke in the big picture of the nation's food system, what can't be quantified in economic terms is their role in raising vital food-related issues. Indeed, CSA's biggest contribution might be its ability to serve as a crucible for a variety of critical problems, from public education to health and nutrition to landownership and agricultural access.

In the end, the term *community-supported agriculture* might not exactly capture the spirit of these pioneering grower–consumer arrangements. The word *supported* is the most objectionable part of the phrase, for it implies that the primary (if not sole) aim of CSAs is to aid farmers. *Shared,* or *stewarded,* or *sustained,* might be more appropriate. According to University of Massachusetts Cooperative Extension educator Cathy Roth, even pioneer Robyn Van En was dissatisfied with the term. "Robyn confided with me toward the end of her life," Ross said, "that she would have been happy with the term 'Agriculturally Supported Communities.'" Farmers benefit, but they do so by sustaining the health of the community at large.

In his 1955 book *Our Vanishing Landscape,* historian Eric Sloane wrote: "A hundred or more years ago, whether you were a blacksmith, a butcher, a carpenter, a politician, or a banker, you were also a farmer. If you were retired, you were a 'gentleman farmer.' Even the earliest silk-hatted and powder-wigged American had gnarled hands that knew the plow and the tricks of building a good stone wall."[3]

Some pine nostalgically for those days. Others feel a void of purpose in their lives as connections to gardening and farmwork are severed. Although many aspects of yesteryear have been wisely forgotten, others are worth keeping and in need of revival. CSAs have by no means solved the myriad problems facing the small, diverse farmer today, but they do offer a mechanism by which issues can be discussed within the broader community. And perhaps for the first time since the days of victory gardens, community-supported agriculture has brought the dinner table—and diners—closer to the soil than they've been in a long time.

Chapter 3

Substance Abuse

Alan Thein Durning

The agrarian agenda calls for work not just in protecting soil and pro-
ducing healthy food but also in finding ways for people to meet their
overall needs with significantly lessened adverse effects on land. This
means coming to grips with the physical stuff of our existence: reducing
what we demand; diminishing the adverse effects of production; length-
ening the useful lives of products by improving their quality; and finding
better ways to reuse, restore, and repair. Like nearly all the necessary
tasks of conservation, however, this work requires communal structures
that support it, sustain it, and over time come to expect it.

Prominent among those working to promote such communal struc-
tures and to encourage local people to improve their resource-use prac-
tices is Alan Thein Durning, founder of a Seattle-based organization
called Northwest Environment Watch. As he reports in the following
selection, even urban dwellers can act on agrarian values in specific,
practical ways. Durning recognizes that conservation has a lot to do with
how one lives, alone and with others, and with the choices one makes
every moment of every day. Knowing that and wanting to share what he
has learned, Durning labors to connect people with the histories and
futures of the products that sustain their lives.

Jens Molbak is an urban miner. He runs a high-tech operation,
employing computerized remote sensors to extract zinc, copper,
and nickel from the landscape. But Jens, whose family has long
been in the Northwest, does not follow buried veins. He mines the

tops of bedroom dressers and the bottoms of kitchen junk drawers. His company, Coinstar Inc., based in Bellevue, Washington, makes computerized coin-sorting machines and installs them at supermarkets. The sorters count change dumped in a hopper and dispense vouchers good for cash at the checkout stand. In 1994, the company's first full year of operation, Coinstar's three-score machines recovered seven hundred tons of pennies and other coins. That was seven hundred tons of copper and other metals that did not have to be mined, smelted, and minted. (Someday, Jens says, "We'd like to shut down the penny mint.") It was also seven hundred tons of coins that no longer cluttered people's living spaces.

Seven hundred tons of metal, in a regional economy that gobbles metal by the mountainside, is hardly worth mentioning. But Jens Molbak is the perfect metaphor for the new economy emerging in the Pacific Northwest, an economy that knows waste is money—money lost, money waiting to be made.

His company's existence, and that of the increasing number of companies making a profit by recycling various "waste" products, is a lucid demonstration of the fact that there is a lot of money lying around. Wasted money. Money in the form of discarded pocket change, and money in other forms: unused nickel, copper, and steel; discarded wood, cardboard, and glass; plastic, paper, and concrete; yard clippings, plaster, and rags; food scraps, methyl bromide, and every other unwanted substance. Even sewage sludge is money in disguise.

A similar idea applies to things—or qualities—that, while not yet discarded, could be designed out of existence. Think of excessive packaging, planned obsolescence, quick-changing fashions, and disposability; these are "waste" because the same end could be achieved with less stuff. Again, ultimately, all waste is money— money lost, money waiting to be made. The trick, as Jens Molbak's coin counter illustrates, is to figure out how to mine the waste and turn bad money into good.

Put it another way. Think of the economy as a giant organism, as does Herman Daly, professor at the University of Maryland and dean of ecological economists. Money is its circulatory system. Money goes round and round, from businesses to households to businesses to households. There is also a digestive system. It is a one-way trip from ingestion to digestion to excretion. What matters to the world outside the economic organism is not the circulation of money but the digestion of resources. What matters is what the organism eats and excretes. Daly calls this combination of input and output "throughput."

Throughput is a useful word. It fills a hole in our language. It is a way to capture in two syllables the entire process from extraction to manufacture to consumption to disposal for everything from virgin timber to tiddlywinks. It encompasses cement, stone, and gravel; fresh water and farm produce; coal, oil, and natural gas; wood, paper, and cardboard; metals, chemicals, and plastics; textiles, rubber, and road salt; and everything else that is animal, vegetable, or mineral.

The sheer quantity of throughput in the Northwest economy is staggering. On average, it amounts to 115 pounds of mineral, agricultural, and forestry products per person per day. That amounts to a lot of stuff.

The trick of sustainable economics, argues Herman Daly, is to keep the circulatory system going round and round while putting the digestive system on a diet. It has got to be quite a diet, considering that billions of poor people around the world need substantially more than they now have, and considering that world population will grow for decades because most of the world's people are very young—even if all couples immediately decide to have no more than two children.

At the Wuppertal Institute in Germany, economists and ecologists have calculated that, for these reasons, the only thing that will suffice to turn humanity off its collision course with planetary ecology is to increase by a factor of ten the efficiency with which industrial countries use natural resources. For each dollar in the Northwest economy's circulatory system, the region needs to send one-tenth as much stuff down the economy's digestive tract. For each published word read, for example, the region must devise ways to use one-tenth the paper, and for every trip taken, the region must learn to use one-tenth as much fuel. The Northwest must do across the board what replacing bulky copper wires with ultralight fiber optic cables has done in telecommunications: improve the quality of service while reducing resource consumption by an order of magnitude. So *sustainability* means achieving a "factor-ten economy": an economy that extracts a better quality of life from a daily resource diet of just 11—rather than 115—pounds per person.

Factor ten may sound impossibly ambitious. In any event, there is room for dickering over numbers. Make different assumptions about population, poverty, economic growth, and the resilience of the biosphere, and you will get to as much as factor one hundred or as little as factor four. For the present generation, however, these arguments hardly matter. What matters is the need to multiply the efficiency with which the economy uses throughput. What matters is the challenge that coin counter Jens Molbak latched onto: how to create what salespeople call *value* through what garbage people call *reduction.*

Other Northwesterners are accepting the challenge too. They are using brains and bytes instead of pounds and gallons. They are sharing and repairing rather than disposing and replacing. They are substituting recycled for virgin, reusable for throwaway, and safe for toxic. They are designing into oblivion gigawatts of power. They are planning for durability rather than obsolescence. Whether it is in feeding themselves, clothing themselves, sheltering themselves, or meeting their other needs, Northwesterners are inventing ways to mine waste and wastefulness. . . .

✤ ✤ ✤

"Cobbler" is an old-fashioned word, connoting a time when kindly old men in leather aprons stooped over their benches to keep their neighbors shod. Perhaps that is why a historian-turned-boot-mender would paint it on his workshop: "Dave Page, Cobbler." Dave, square-faced and built like a wrestler, explains in soft, crisp sentences, "It is from Old English. It means someone who puts things together so they work." He glances forward briefly to signal that he is done speaking. Then he returns to scanning the shoes piled on tall shelves above his head.

The Northwest's premier footwear recycler is modest about himself but proud of his workers. Seven of them join him each day in a sunlit Seattle storefront to resole mountaineering boots, restitch street shoes, and refinish sandals.

"That's Giuseppe there," Dave says, studying the deft movements of the eldest, a silver-haired Italian who looks a little like Pinocchio's father. Shaking his head in admiration, Dave whispers, "He's a genius. There's nothing he can't do. He's been doing this for forty-seven years. He started as a boy in an Italian boot factory."

Quality, experience, and smart positioning explain Dave's consistent success during a quarter century in which most shoe repair shops—and repair industries generally—have done lackluster business. In 1993, just 117 people earned their livelihoods repairing and shining shoes in the state of Washington, down from 207 forty years earlier. But since 1990, Dave Page, Cobbler, has been growing by more than 25 percent a year, even while the shoe repair industry overall has been stagnant.

There was a time when cobblers, tailors, tinkers, and other fixers of things made up a good share of the Northwest work force—and mending things took up a piece of most people's weeks. But cheap raw materials, mass production, and, later, offshore manufacturing shifted the balance. "New shoes from Brazil and the Orient are so cheap!" Dave says, examining a new heel on a dress boot. "Go to these discount shoe stores and you can get two pairs of decent-looking shoes for eighty dollars. It would cost more to resole them than to replace them." So the worn shoes, with all the energy, materials, and workmanship that went into them, become dump filler.

The cost of repair commonly exceeds the cost of replacement not only for shoes but for all kinds of manufactured goods: blue jeans, radios, toys, answering machines, dishes, computer screens, auto parts, telephones, books, and even the hand tools used to repair things. The repair industries generally have been in relative decline. In Oregon, for example, the number of jobs repairing things has not kept pace with overall employment; between 1970 and 1992, total employment grew one-third faster than employment in nonautomotive repair services.

Compounding the trend away from repair has been the relentless, marketing-driven expansion of fashion. Two centuries ago, only the rich worried about being in style, and then only for their dress clothing. Today, all classes are afflicted, and for everything from wristwatches to automobiles. Fashion is an insidious form of planned obsolescence: things become useless long before they wear out.

In the twenty-six years he has been in business, Dave has watched fashion overrun first the athletic shoe trade, then the ski boot trade, and finally the trail boot trade. As fashion advanced, repair—and durability—retreated. And shoe designers stopped paying attention to whether a shoe could be fixed. It no longer mattered.

Even in mountaineering boots, a realm so far untouched by fashion, Dave has noticed a decline in durability. "We've resoled some boots from the fifties and sixties five and six times." New boots are lighter and give better performance, but "they don't last as long."

The cost advantages of mass production will not go away, but in a factor-ten economy, designers of mass-produced goods would put as much emphasis on durability and repairability as on style. And cobblers and other repair industries would flourish.

✤ ✤ ✤

Dave Page, fifty-six years old, was born in northern Wisconsin and came to the Northwest to run the mile for the University of Washington. By the time he had finished his Ph.D. in history, however, he was more interested in vertical miles: he was an avid mountaineer, captivated by the rock and ice of the Cascades. "At the time, only one guy in the country was resoling mountain boots. He was in Colorado. I got tired of waiting the six months it took him to send the boots back."

Dave taught himself cobbling, mostly by ripping apart old boots, and was immediately overwhelmed by climbers wanting repair work done "before next weekend." So he quit his job lecturing to undergraduates on American history and spent five years in his basement perfecting his craft. Word of mouth spread his name to climbers all over the continent. Convoys of UPS trucks began rolling up to his house.

Only years later, after he had hired others and moved to successively larger shops, did he expand from boots to other kinds of footwear. "At first we only did it as a favor to the neighborhood. Now we get shoes from all over the city." There are few other places left to go. Most shoe repair shops, one- and two-worker outfits already buffeted by dwindling business, have been unable to keep up with shoemaking technology. Asian manufacturers, for example, use thermoplastic shoe bottoms—consisting of the sole and what it sticks to—that require expensive equipment and advanced adhesives to replace.

Dave grows somber, thrusting out his jaw slightly and lowering his head of slate-gray hair. Scores of good cobblers have given up, he says, selling out to "guys with two weeks of training who have always wanted to own a business. The skill level is way off."

Repair is to recycling as recycling is to dumping: another step towards factor ten. Waste recycling in the Northwest is surging, but repair shops are just beginning to revive, as are other institutionalized forms of reuse—lending libraries, remanufacturing plants, rental outfits, and thrift stores. The integration of reuse into product manufacturing, moreover, has yet to commence in earnest. Even Northwesterners rarely design things to endure, to foster repair, or to accept upgrades as technology improves.

A handful of remanufacturing plants such as GreenDisk buy used manufactured goods in bulk and refurbish them for resale. From its headquarters on the outskirts of greater Seattle, GreenDisk collects computer disks by the truckload and restores them to new condition. Other Northwest companies do the same for telephones, auto tires and parts, electrical equipment, office furniture, computers, machine tools, and laser-printer toner cartridges. Again, remanufacturing works best when products are designed for continuous restoration.

The secondhand trade has experienced quick growth since the late 1980s. Between 1989 and 1994, thrift stores in Idaho, Oregon, and Washington expanded their payrolls by half, to more than five thousand workers at one thousand establishments. Yet these figures may reflect the acceleration of fashion rather than a shift to reduction. In Washington, for example, thrift-store employment has grown thirtyfold since consumerism became the norm in the 1950s. There is just so much new stuff around now that more of it goes off to Goodwill.

In contrast to the rise of secondhand shops generally, used-book stores have been dropping like flies. The advent of discount, volume bookstores did in ten secondhand booksellers in Vancouver in 1994 and early 1995 alone.

Unmeasured in business statistics is the shadow economy of yard sales, rummage sales, church sales, and swap meets. Possibly larger as a distributor of manufactured goods is the bustle of sharing, loaning, and borrowing between friends and neighbors. Again, no surveys or statistics track this sector of the economy, although most observers of community life believe that these nonmarket channels of trade have atrophied. Frontier towns and early-century neighborhoods were thoroughly bound in the mutual reciprocity of borrowing and lending. With affluence has come independence: a Northwest homeowner who finds herself in need of a C-clamp one Saturday is as likely to drive fifteen miles to Home Depot as to ring doorbells around her cul-de-sac until she finds one.

Still, Northwesterners do share, sometimes with remarkable creativity. At the Evergreen State College in Olympia, Washington, for example, the student union building is home to a free box where unwanted clothes are left for others to pick over. Entering students often deposit there the new clothes with which their parents have dutifully outfitted them; then they garb themselves in the tattered jeans and T-shirts left by departing seniors. Meanwhile, graduates heading to job interviews pick out the new apparel supplied by the new students, and the symbiosis is complete.

Lending libraries perform the same function in a more formal way: they allow sharing on a countywide scale for books, periodicals, sound recordings, and videotapes. Typical Northwesterners checked out ten books and other items in 1992, up from six in 1970. Of course, many people are not in the habit of using libraries at all; most of the 114 million items loaned in 1992 went to a core of dedicated patrons.

In one Northwest library or another, you can check out board games, cameras, CD-ROMs, computer software, electric engravers, framed paintings, hand tools, movie projectors, puppets, sheet music, tape recorders, and video games. You can also surf the Internet or practice the piano. In Portland, Oregon, Michael Keating of the United Community Action Network created a bicycle library of the streets. He repaired and repainted one hundred junked two-wheelers and fitted them with these instructions: "Free community bike. Please return to a major street for others to reuse. Use at your own risk." Weeks later, thirty-two blue painted community bikes hit the streets of Victoria as well, and another score showed up in Salem, Oregon.

The for-profit version of the library is the rental center. The Northwest has long had more than a thousand firms that earn their keep by keeping their merchandise. They rent out tools, tuxedos, wedding gowns, party supplies, medical equipment, trucks, skis, beds, furniture, dining chairs, tablecloths, roller blades, place settings, appliances, sound systems, photo equipment, diapers, musical instruments, heavy machinery, costumes, forklifts, boats, bikes, televisions, computers, and portable toilets. Video rental is the comer in this crowd: it grew by half between 1989 and 1993 in Washington, and it employed almost as many Oregonians in 1992 as did the mining industry.

In the end, however, every variety of reuse is limited by the rapid obsolescence designed into so-called durable goods. Consider household appliances. The Washington Department of Community, Trade, and Economic Development brags that recyclers collected, shredded, sorted, and melted down 126,000 tons of household appliances in 1992. What a waste! Most new dishwashers bought in the Northwest are designed to last just ten years. Refrigerators, washers, and dryers last thirteen or fourteen years, and ranges last seventeen. These appliances are cheaper, lighter, more energy efficient, and do their job better than the models they replace, but they do not last any longer; and, because they are assembled, without bolts, from molded plastic parts, they are more difficult to repair or disassemble for scrap. In a factor-ten economy, appliances would be designed to last longer and to be modular enough that repair, remanufacture, and upgrading would be the rule, not the exception. . . .

❋ ❋ ❋

An Old World steamer trunk sits in the corner of Steve and Kris Loken's living room. It is handmade from aged scraps of salvaged pine. The scraps are held together by wooden pegs. The corners are edged in steel, and the whole thing is lovingly painted. In it, Steve's grandparents brought their belongings from Norway to the town of Heartland, Minnesota. Now the trunk looks out over Missoula,

Montana, where Steve has lived since his microbus broke down there in 1969. The trunk sits in the living room for three reasons. The first is obvious; the second and third take some explaining.

The first reason is that the trunk, most of a century old, is a testament to the Loken family's tradition of crafting things to last and then taking care of them. The house in which it sits is a testament to the vitality of that tradition in Steve, who is a builder. The house is a stunning leap toward factor ten: it was made with few virgin resources, it was made for durability, and it was made to consume few resources during its operating life. Steve had low throughput firmly in mind from start to finish.

Steve had another thing in mind, too. He wanted to "hit the soft, paunchy underbelly" of the Northwest's house-buying public. In the Northwest, builders broke ground on 126,440 units of new housing in 1993. Most of these were not resource-thrifty apartments or rowhouses in town. They were oversized, split-level ranch houses with two-car garages on the rural fringe. They were black holes of throughput: small but with unbelievable gravitational pull, able to suck resources from great distances. They were, in Steve's words, "the underbelly."

Buildings are secret leaders in the throughput leagues. Construction accounts for 40 percent of the raw materials consumed by the U.S. economy, according to David Morris of the Institute for Local Self-Reliance in Minneapolis. Northwest buildings use a third of all the Northwest's energy, mostly for light and heat, and manufacturing building components takes up almost another tenth. Construction accounts for nearly half of U.S. copper consumption, two-fifths of wood consumption, and an even larger share of old-growth lumber consumption because old-growth is big enough and strong enough to yield structural beams.

An average new house contains ten thousand board feet of framing lumber; thirteen thousand square feet of gypsum wallboard, plywood sheathing, and exterior siding; fifty cubic yards of concrete; four hundred linear feet of copper and plastic pipe; fifty gallons of paint; and three hundred pounds of nails, according to the National Association of Home Builders.

All together, a typical U.S. house weighs in at 150 metric tons. Demolishing it generates as much waste as its inhabitants would have set out in the rubbish bin during most of its eighty-year life span. When a new house is built to replace it, another seven tons of waste are carted to the dump. In 1993, almost one-fourth of Portland, Oregon's solid-waste stream was from construction and demolition.

Conscious of such effects, Steve set out to erect something that smelled and tasted like the underbelly but lacked the gravity. He wanted to show that reduction could look like the ads in *Town and Country*. So he built an oversized split-level ranch house with a double garage on the rural fringe. But he made it with one-fifth as much virgin wood, a similarly shrunken percentage of most other virgin materials, and a lot less stuff overall: the building weighs in as a lightweight compared to most houses of its size.

The house—which, for all its innovations, cost just 15 percent extra—also outperforms conventional houses in other ways. Despite its draftproof construction, air quality indoors is superb. Despite its light weight, it will likely last three times longer than an ordinary house. It holds heat and coolness six times better than Montana's typical new house, and it is three times as frugal with water. During construction in 1990, Steve and his crew minimized and recycled waste down to a few Dumpster loads.

The Faustian bargain for Steve Loken, who came to building after studying wildlife biology and living briefly on a commune, was that the only way he could jab at the underbelly was to become part of it himself. To get a loan to build the house, he had to promise to live in it, precisely what he did not want to do, preferring a community-oriented urban setting where his two kids would be able to grow up unchauffeured. The banks, however, would finance the project only if it was going to be "owner occupied."

Beyond requiring that Steve buy the house himself, the bank also stipulated that it be big—to guarantee that appraised value, an appraiser's estimate of its resale price, would cover the loan. Never mind what Steve was trying to demonstrate or what Steve and Kris wanted as a place to raise their two children, both adopted from Korea: the appraisers said big houses hold value better than small ones.

Looking out across the pines and grasses of the northern Rockies, Steve shakes his head. "The best living room in the world is just outside your threshold, so why try to replicate that? Just go outside." The median house in Montana is 1,300 square feet, smaller by a quarter than the national average. Steve's house, thanks to the bank, is a whopping 3,000 square feet.

Now, he and Kris are stuck in it. "It's too big. It's not us. Since we moved in here, real estate prices have gone so high I can't afford to stay here or buy anything else either. I'm trapped. It's really bizarre," he says, speaking with quiet rage.

This rage is the second reason for the trunk's prominent placement: it's there ready to pack. It's a statement of intent. . . .

❊   ❊   ❊

In small part due to the house, and in somewhat larger part due to Steve, the building industry in the Northwest has changed quickly since he finished his house in 1990. Other demonstration homes have gone up in Vancouver and Portland. More than a score of stores have opened to buy and sell salvaged building components. Newsletters and conferences have proliferated. Composite materials have swept into the market, shepherded by steep increases in the price of virgin lumber. Portland has aggressively promoted construction-site recycling, quickly diverting half of the construction waste stream away from city dumps. Steve is pleased, but far from satisfied.

❊　❊　❊

Maybe that is because so few visitors, builders, or others pay any attention to the third reason that the old steamer trunk is waiting in the corner—but, again, that reason takes some explaining.

Steve was not born into the Norwegian family who brought that trunk to Heartland, Minnesota. "My birth father was a Swiss Catholic, and my mother was a German-Austrian Lutheran." His steely blue eyes and sharp, ruddy nose would fit in well in the Alps. "That was not a cool thing—for Lutherans and Catholics to be cohabitating. They were in their early twenties. Met at a barn dance." Steve was put up for adoption.

From his adoptive parents, and his father in particular, Steve got his intense "materialism"—his devotion to materials, their properties, potentialities, and functions, their intrinsic stuffness. Steve also learned the values of frugality and thrift that go with this true form of materialism. To Steve's mind, the problem is not that Northwesterners are too materialistic but that they are not materialistic enough. Northwesterners care *about* things without caring *for* them. They seek the forms of things rather than their substances; they are, in a way, substance abusers, treating all ailments with infusions of new objects.

"When I was a kid, we went to the dump every weekend. We always left with more stuff than we came with." Steve's son Rye is climbing over him as he talks.

"Later, my dad worked for the highway department. When I-90 was going through, we went out for picnics after work, at night, and on weekends. And we would go out with crowbars, hammers, and go to places he knew were going to be bulldozed or burned in a few weeks." Rye is resting his head on his father's shoulder. It's getting late.

"We would take these buildings apart and salvage everything that we could. I had great fun unbuilding when I was a kid. We salvaged everything and stored it. On our farm we built our house, garage, and outbuildings as we could afford them." Steve stands up and rocks Rye slowly on his shoulder.

"We started out living in a basement house for six or seven years, a basement with tar paper and mineral-roll roofing. We lived there and continued to build until we finished the house. It was kind of a grow-home. We all got a chance to build and work on it. That's how I learned construction and carpentry. We built our house, three-quarters of it, with salvaged material. Back then it was really excellent wood. And we built that house to last.

"But look," he says, "here's the question: permanency. We have built permanent and stout even though we have a real impermanent, transient population."

Steve's daughter Kyra comes out of her room to get a goodnight kiss. She hugs his leg and lingers to listen. "In terms of its resource consumption, the renovation industry is rapidly approaching what we use in new construction." Building renovations are projected to surpass new construction as a source of income for builders by the year 2010 in the United States as a whole, and somewhat later in the young, fast-growing Northwest. Much remodeling is motivated by concerns about fashion rather than performance: Formica countertops from the seventies that have another thirty years of useful life in them are ripped out and landfilled to make way for mock granite that will, by all odds, become trash too, when the next owners come along.

"Maybe modular, lightweight, and movable is the way to go," suggests Steve. "My grandparents' trunk carried everything they needed. That's my ideal."

And that's the third reason for the trunk's place of honor in the living room. It was enough for them, and it was built to fit their migratory predilections.

In at least one small, secretive way, Steve tried to build portability into his house as well. In so doing, he committed an act almost unimaginably subversive in the building trades. He fastened much of his house together with screws rather than nails. "Oh, the house could last 200 or 250 years. But I want my kids to be able to take it apart and move it. They're not going to be able to live up here. This is crazy."

Chapter 4

Prairie University

Stephanie Mills

One of the much-needed conservation tasks of the new century is to restore health to damaged landscapes while working to defend those that retain their diversity and vigor. Restoration work often requires professional expertise, yet it is work that can engage—and very much needs to engage—the energy, vision, and passion of ordinary people everywhere, urban and rural. This selection describes successful restoration work being undertaken in and around one of the country's largest cities, Chicago. That work, bringing life back to prairies, savannas, and wetlands, has helped heal not just particular publicly owned tracts and surrounding acres but also the people involved in the restoration work and the social communities they help form. They have gained health, individually and collectively, by strengthening their physical and emotional bonds to the land and to one another.

From her home in northern Michigan, Stephanie Mills has worked for more than twenty-five years to promote conservation and social change across the land. In this selection, she exalts and encourages one of the many forms of land-focused local action that fit within the agrarian cause.

Late in the 1990s, the restoration work that Mills describes encountered resistance, both from neighbors unfamiliar with the underlying ecological principles and from animal welfare advocates concerned about deer-culling efforts. By 2000, however, interim restrictions on restoration work had largely been lifted.

"I want it to be next spring already," says John Balaban. Balaban is one of a network of 3,800 volunteer stewards, laypeople who donate

45

their time to the work of restoring prairie and savanna ecosystems in northern Illinoisan nature preserves. Many of these preserves are suburban and cluster around the North Branch of the Chicago River, within fifteen to twenty miles from the downtown Loop. At the height of summer, Balaban wants it to be "next spring already" because with every passing year, more is learned about the art and science of repairing damaged landscapes, and for many years John Balaban and Jane, his wife, have been ace practitioners.

Jane Balaban, a hospital administrator, is bright with enthusiasm for stewardship and restoration as she welcomes me to her comfortable home and offers me some lunch. It's August 1991. I have come here to rendezvous with Steve Packard, Science Director of the Illinois chapter of The Nature Conservancy and a prime mover of the North Branch Prairie Restoration Project and the Volunteer Stewardship Network.[1] He has kindly consented to take me on a field trip to some of the sites the volunteers are working on within the Forest Preserve District of Cook County, 67,000 acres of mostly wild land scattered north and westward through the Chicago metropolis. (Despite the sylvan name, the ecosystems contained in these preserves are not only forests, but grasslands and savannas, which the early settlers called oak openings.)

Fire is the elemental fact in the evolution and dynamic equilibrium of these landscapes, as it is in . . . all grassland ecosystems. Left to its own vegetal devices brush would shade out the oaks and the many native grasses, sedges, rushes, and forbs that characterize these rich tapestries of life in which no single color or thread, no one species, predominates.

The midwestern tallgrass prairie all but vanished, overturned by the steel-bladed plow in favor of the relentless monotony of agriculture, which amounts to planting lots of the same damn thing— corn in Illinois, usually—and then attempting to protect it from all the plagues that even-aged monocultures are prey to. Sodbusters and years later the advance of suburbia made wildfires, and hence the prairie and oak openings that depended on them, even more a thing of the past. Benign neglect will not suffice to keep these precious few remnant ecosystems alive. Fires must be lit and skillfully managed, and many other tasks performed as well.

Prairie and savanna restoration work is terrifically labor-intensive, requiring thousands of hours of work to remove exotic species of plants from the areas to be restored, to lop the light-hogging buckthorn and ash saplings, to scythe weeds, to burn the leaf litter and grasses, and to gather, thresh, label, store, and then sow and rake seed from hundreds of different varieties of rare plants. Volunteers are *sine qua non.*

The "idea that these sites can't exist without our help anymore"—as John Balaban put it, "you can't just preserve something by building a fence around it," because of "how dependent that structure [of the ecosystem] is on our interference"—convinced the Balabans of the need for their stewardship as it has hundreds of other volunteers.

The North Branch Prairie Restoration Project is a remarkable mobilization of human energy and is instructing a sizable group of amateurs in the Chicago megalopolis in some of the finer points of prairie and savanna ecology, preserve management, field botany, plant conservation, and horticultural technique. To enhance and embellish this learning, volunteers coordinated by Laurel Ross, who like Packard works for the Illinois chapter of The Nature Conservancy, publish a thrice-yearly catalogue for "Prairie University." This university-without-walls has a curriculum consisting of courses available at educational institutions throughout the region as well as workshops, seminars, and field trips offered by various museums and learned societies and some by the volunteers themselves. Among the scores of offerings listed are: aerial photography, art/nature, biology, birds, botany, chemistry of the environment, controlled burns, ecology, endangered species, entomology, environmental science, evolution, field methods, gardening and landscaping, geography, geology, mapping, mushrooms, trees, wetlands, and zoology. The City of Chicago's motto is *"Urbs in Horto."*

The Prairie University motto is *"Discere in Horto."* It means "learn in the garden" (also, but unintentionally, "learn by exhorting").[2] Paging through the Prairie University catalogue is a cheering experience, evidence of the hankering of so many members of the human community to exchange knowledge of natural history and promote ecological literacy.

"Know the plants" may be coequal with the injunction to know thyself as a fundamental responsibility of *Homo sapiens.* Plants tell us where we are, are the basis of our sustenance and our atmosphere. They make the life of the land. Ecological restoration is very much about the reinstatement of native plant populations and communities, and letting loose the processes they require for their evolution.

It is also important, in becoming native again, ourselves, to know the animals. And, since *Homo sapiens* are calling the shots, working with the dynamics of our own species is a fundamental aspect of restoration. In addition to all the botanical and occasional entomological savvy that informs the work on the North Branch Prairie Restoration Project, there is a good deal of human understanding. Not for nothing was Steve Packard a student of social anthropology. He and his prairie project colleagues work in an urban setting where people are overwhelmingly the dominant species.

What Steve Packard is famous, even notorious (among pedigreed plant systematists), for is his assertion that there is a distinct but forgotten community of grasses, shrubs, and wildflowers existing in oak savanna. Before Packard's savanna hypothesis, the idea was, roughly, that what hadn't been oak forest had been tallgrass prairie. Therefore prairie was what it would make sense to restore to in these open woodland places. The problem was that the prairie restoration didn't take. In areas of the Cook County forest preserves where the great old oaks lingered, even after clearing out the underbrush and planting prairie species in the newly opened but partly shaded ground, volunteers observed that the presumed natives failed to thrive. The prairie plants couldn't seem to take hold and displace the invading woody species. At the same time, what Packard referred to as "a few oddball species of plants" characteristic of neither prairie nor forest kept popping up in the sites.

By some masterful sleuthing Steve Packard began to speculate about what community of vegetation might have flourished originally under scattered tree canopy. Through taxonomic and historical research, he assembled the identities of the few "oddball species" and many other plants in the oak savanna complex. Eventually he came up with a list of plants that turned out to be comprised of woodland, not prairie, grasses, and many herbs. Scouts began to locate other small remnants of savanna communities within the vicinity of the preserves and to gather seed and propagate plants for savanna restoration work.

Before we set out for the Somme Woods Preserve, northernmost of the North Branch project sites, and Packard's "favorite place in the world," he took me around to the side of the Balabans' house to show me a little garden where the couple was growing some of the rare prairie and savanna plants that they and other volunteers are restoring to these sites. Household cultivation of these scarce plants weaves humans together with nature, and the developed landscape with the reflourishing landscape of the preserves.

Over the long run this learning in restoration—along with the landscaping-with-native-plants practice increasingly promoted by garden clubs, enlightened landscape architects across the country, and networkers like Louise Lacey (editor and publisher of the northern Californian *Growing Native*)—could enliven, maybe even transform, the biological monotony and heterogeneity of human settlements.[3] Such work, and the propagation going on in the Balabans' garden, and hundreds like it, is clearly reinhabitory. . . .

❧ ❧ ❧

Steve Packard apparently was born to be an ecological restorationist. Just as some kids are always going to be firefighters or doctors or *danseurs,* he was always going to be a restorationist, although it took him a while to discover how that was the case. Since junior high school he has found ways to see places become more like themselves, to regain their true identity through time.

"I used to make terrariums when I was a kid," Packard said later in our encounter. He told stories of leading his teachers out through the Massachusetts woods in the winter and dowsing for specific plants under the snow, and then going on to teach others about making terraria. "The more germane thing," Packard later added, "is that I planted native strains of pines and hemlocks in my behind-the-backyard 'bird sanctuary,' fiercely resisting my dad's pressure to use the 'improved' strains you can buy in a nursery."

Highly educated (Harvard, but reticent about it), Steve Packard now goes about retrieving lost landscapes through dirty-fingernails work mixed with long-term contemplation and cogitation, as absorbed as a chess master. During my day visiting restoration sites with him and later over dinner, Packard's conversation evinced a curious amalgam of self-effacement and pride. He credits his initial learning of plants to Roger Tory Peterson and Margaret McKenny's *A Field Guide to Wildflowers of Northeastern and North-central North America.* The book, Packard says, put in the layperson's hands a tool for calling the plants by their proper names without having to do the taxonomic trudging. (He has long since learned to wield a mean taxonomic key, sussing out even such cryptic native flora as grasses and sedges.)

He had a botanical epiphany in Chicago in 1975 when he was transfixed by the sight of a beautiful, fresh white flower, *Lychnis alba,* the evening campion, growing on some desolate ground behind a factory. At about the same time he read that someone had discovered the last prairie. "When I traveled down to see it, it seemed pretty magical in some ways . . . in some ways it seemed like a lot of weeds." The magic of it won out and thus began a vocation in prairie and savanna restoration—one that over time would come to entail some great battles with a lot of weeds.

❧   ❧   ❧

The visionary prairie-phile landscape architect Jens Jensen played a leading role in the establishment of the Forest Preserve District of Cook County. Jensen practiced in the region around Chicago in the early twentieth century. In addition to being one of the first to celebrate native ecosystems in public parks, Jensen promoted volunteer participation in the work on certain of his projects. But neglect of the necessary management practices led to the preserves' degeneration. Enter the North Branch Prairie Restoration Project, whose *raison d'être,* according to its Mission Statement, is "to assist the Forest Preserve District of Cook County and other agencies in protecting and restoring native Illinois ecosystems."

The preserves were established sixty years before the Volunteer Stewardship Network got started. There have been no fires in them since the mid-nineteenth century. In their ecologically degenerate state, much of the land in the preserves had been abused and used the way vacant lots generally are—as kids' rendezvous, party spots, offhand trash heaps. Removal of soggy car seats was an early step in the restoration process. The absence of fire had allowed a proliferation of brush, particularly the European buckthorn, which shades out the oak seedlings and stifles the lower branches of the centuries-old oaks that remain, talismans of a time before Chicago's settlement by Europeans. The restoration work on these preserves includes bringing the brush to heel, then planting seeds of the native climax vegetation. The theory is that the former climax community will succeed, in both the vernacular and ecological senses of that term. A climax community maintains conditions—soil texture and nutrients, shade, and biotic richness—that favor its continuation. Barring disturbance, climax communities are very stable through time. Nature being the prankster she is, however, disturbance inevitably occurs, and in the disturbed areas successional processes are set in motion. Different groups of plants are adapted to take advantage of the set of conditions prevalent at the moment, each through its life cycle changing those conditions toward climax.

In the succession to forest, for instance, the movement is from open, sun-drenched poor soil through ephemeral light-hungry species to long-lived, shade-tolerant trees that will form a canopy with moderated conditions of temperature and humidity below. Grassland communities and oak savannas seem to create the conditions required to invite periodic burns which kill the fire-intolerant brush species that would shade out their light-loving members. Notes Packard: "Prairie, savanna, and oak woodland might all be called 'fire climax' communities. Many other ancient communities are also known now to require occasional disturbance (fire, flood, blow-down, disease). Natural disturbance functions differently from the catastrophic disturbance that humans often wreak."

From the roadway, there's no clue to the extraordinary botanical goings-on at the Somme Woods Preserve. The crafty stewards have left a hedge of buckthorn around the perimeter of the woods, an impenetrable barrier to deter the insensitive, or the destructive visitor (the sort who, until recently, had come to drink and neck and hare around on dirt bikes). On arrival we threaded through a small opening. Packard charged me to behold "a sick, miserable ecosystem that used to be oak savanna." At the beginning this place must have looked horribly unpromising. "There were pockets of good stuff and dribs and drabs of this and that species, in dwindling numbers, dodging the disturbances," notes Packard. "Or perhaps even dependent on those disturbances—in the absence of fire." He pointed to a stand of locust trees that the volunteers had girdled, killing them to let the light shine on the tree species that belong there—bur oaks and white oaks. He pointed out a young oak that was saved from dark demise under a dense canopy. He explained the hallmarks of fire-adapted trees: their corky bark, their fat twigs, their ability to resprout.

Our passing from dark to light to dappled light was the subtext for the lesson that Packard imparted as we walked. We moved first through dark, sterile thickets, with their bare, caked mud floors, then prairies, and then the oak savannas that Steve Packard considers the epitome of desirable terrain. The contrast between the richness of the communities that the network has reestablished in the savannas and the nullity of the ground under the buckthorn is truly dramatic. I had no idea there were so many different kinds of grasses. In the course of our walk we remarked on perhaps two dozen different species and many sedges as well. . . .

❧   ❧   ❧

The thing that is changed forever (or for a long time into the future, anyway) about the life of this renewed ecosystem is that the fire it needs must be set and tended by knowledgeable humans. It calls for modern urban slash-and-burn horticulture. The volunteers cut out the brush to roll back the shade. Once there's enough fuel on the ground—leaf litter from the native plants—then fires can burn now and again and maintain the vegetation complex. Prior to settlement most of these fires were set by indigenous people to maintain clearings for game. Other wildfires would have been lightning-set. The occurrence and intensity of the fires were as irregular as rainfall. Different fires burn at different temperatures; different plants succumb to different fires burning at different temperatures. The fact that some plants survive and some don't results in a mosaic pattern in the vegetation rather than in a wall-to-wall carpet of repetitious motif.

There was a moment of pure bliss for me in this postage stamp of healthy land. We walked in an island of wildness that had a surf of automobile traffic noise beating around its shores. But there were enough cicadas stridulating to give the traffic some stiff competition. The weather couldn't have been improved on, the sky blue and bright with high clouds, twinkling with butterflies. Butterflies like these places. Packard remarked that even if you didn't like plants, this work would be worth it for the butterflies. We watched a tiger swallowtail and an ebony-colored black swallowtail, or it may have been a mourning cloak of similar size, dancing a *pas de deux,* or they may have been contending for territory across a prairie opening. There was just something about the light. And about the excitement of being among so many different kinds of plants, of feasting the eye and the subtler senses on bountiful detail. Always something fresh to point to and ask, "What's that?" It was a great privilege to be with someone who knew the answer, who may in fact have caused the cluster of false indigo or Indian grass to be growing there at all.

The Somme Woods site which we toured is, as is any restoration site, a lab. It's a story featuring the uneasy waltz of Science and Nature (Science all too often showing up for the dance in steel-toed work boots while Nature goes barefoot). Restoration ecology is experimental science, a science of love and altruism. In its attempts to reverse the processes of ecosystem degradation it runs exactly counter to the market system, to land speculation, to the whole cultural attitude of regarding the Earth as commodity rather than community. It is a soft-souled science.

At one point in the ramble I asked Packard what he felt in such places. If he'd been there by himself what would he be feeling? Would he feel a sense of satisfaction? After some thought he said he was wary of taking personal credit for this. He likes almost to forget that he and others have had anything to do with this place, this recovering ecosystem. It is as it should be, and that's the point. He talked about the satisfaction that comes of being a part of something good and useful. He made an analogy with taking a child to do something fun. The child mayn't necessarily thank you, but being able to witness his or her enjoyment is reward enough.

Packard talks about organizing the "generous impulses that people have toward their world." As an organizer his method is to say, "Here's the necessary support, here's the necessary freedom and authority: go ahead." It amounts to a deft liberation of energy, akin to the emancipation that the burning and girdling of underbrush affords the hitherto scarce plants like the sweet black-eyed Susan which, said Packard, "needed the same release that the people needed."

An ecological community is founded on plants, the primary producers, with intertwined loops of consumers—herbivores, carnivores, and omnivores—relating hungrily in the midst of the vegetation. An ultimate index of health is the presence of big, wide-ranging predators. Because the preserves are so constricted and isolated in size, the likelihood of reinstating larger creatures seems slight just now. But very large creatures once were integral to these ecosystems. Chicago was a place where the buffalo roamed and what became Vincennes Avenue was once a bison track. However, the Somme Woods Preserve, as Packard told me that day, is host to red-tailed hawk, sparrow hawk, and the rare Cooper's hawk. Evidence that a goshawk wintered here is provided by the piles of pigeon feathers found all over the preserve. So there are avian predators and coyotes that have returned, but no wolves or bison are expected for a while. The question is whether these restorations will be sustainable indefinitely in absence of the larger animals that were once a part of these communities. On the savannas and prairies humans have to stand in now for the forces of Nature, to play the part of lightning storms, of buffalo, and of wolves and cougars, as they cull the deer.

White-tailed deer have become so numerous in the North Branch prairies that their grazing threatens to do irreversible harm to these painstakingly restored ecosystems. The decision to kill as many of the deer as possible triggered a controversy among the Chicagoans paying attention to such things and prompted some opposition by animal-rights activists. Laurel Ross, who coordinates the Volunteer Stewardship Network, engaged the subject in her writings:

> These caring and well-meaning people speak loudly and passionately against the culling of deer, as if saving some individuals should take precedence over saving the precious and irreplaceable system which supports them. It is important for us not to lose sight

of the fact that *we* are animal rights activists. We put
our time and our energy into restoring and preserving
habitat so that hundreds of species of animals and
plants may thrive.

When newcomers to the Prairie Project challenge
the seemingly brutal methods used to control brush
and pest plant species, it is explained how and why
this is critical to the recovery of an ailing ecosystem.

Deer control parallels other management tech-
niques. Because we know fire is essential but cannot
live with uncontrolled wild fire in a populated area,
we have substituted safe, controlled burning. Nature
provided wolves, bears, mountain lions and people to
keep deer populations in balance. If we cannot tolerate these large predators in our cities, then we owe
our natural areas a substitute they can live with.[4]

So how does the Forest Preserve District actually do the dire
deed? A trench is dug to hide a net. Bait is set out for the deer.
When they are gathered in sufficient numbers, they trigger the
flinging, by small rockets, of a net over the herd. Then the deer are
shot point-blank. Their meat is distributed to various charities.

Counterbalancing the drama of sweat, smoke, and blood being
shed in controlling superabundant species, an aspect of savanna
restoration, which is in some ways the essence of the whole thing, is
seed collection. There's no catalog whence you can order seed
envelopes for the scores of woodland species the project seeks to
reestablish in their rightful venues. (Besides, even if such could be
purchased, it would lack the critical attribute of ecotypy—genetic
adaptation to a certain locale.)

Seed collection is the endeavor Laurel Ross superintends as a
volunteer. On a visit in the fall of 1992 to a different restoration site
in the forest preserves—the Bunker Hill Prairie—I tagged along
during a field exercise. During this consultation with restorationist
Tom Vanderpoel, Ross was also reflexively scouting seeds. As we

walked along in the company of John and Jane Balaban and several other volunteers interested in refining their understanding of, and planning for, the ongoing restoration and management of this preserve, Ross noticed smilax seeds ripening. In another place, she noticed that a certain sedge's seeds were ripe and produced a half-dozen plastic bags into which we were directed to gather the harvest.

Plants are impressively protean, taking many different forms, shapes, and sizes in the course of their lives. Consider, for instance, the difference between an acorn and an oak. An untrained observer (such as myself) may see the same plant in a half a dozen different places, having made its adaptation to a half a dozen different sets of circumstances—to peculiarities of microhabitat or to passage of time—and be unable to perceive that they're all individuals of the same species. Thus the ability to recognize hundreds of different plant species, and in the various stages of their life cycle, is real *savoir vert*.

"It's like people," Ross says. "You have to know them really well, so that you can know them walking away from you, when they've gained ten pounds, when they're half asleep." ...

❈　❈　❈

Laurel Ross's transitive goodness has to do with connection—her connection with the people working as stewards on the prairie and savanna restoration projects dotted throughout Chicago; and the connections she effects between people and the communities of flora whose revival is the network's reason for being. One gets the impression that she and her colleague Steve Packard are never not busier than mere mortals could, or should, be. Their projects, wildly successful in many respects, are extensive (27,364 acres, 207 sites) and intensive (about one steward per site, 3,809 volunteers, over 50,000 hours worked in 1992). It's also abundantly clear that these people love their work, thrive in it to an uncommon degree. Ross exhibits an appealing combination of sensitivity and practicality, and general wisdom, sources for that latter quality likely being her engagement with its two biggest wellsprings, Nature and Culture. Hope does not seem to be a problem for her. Her stories of restoration endeavors and volunteers are colored by delight. No misanthropy in this kid. In February 1993, downtown in Chicago, Ross spent an hour in conversation with me, speaking of the rewards in the work.

❖   ❖   ❖

"The reason that people are so willing to work so hard in the Volunteer Stewardship Network," says Laurel Ross, "is that they are looking for something meaningful to do. People think of raising their children as important. People think of making art as important. This is right up there. It's more important to a lot of people than their jobs.

"So many of the life forms we're dealing with are in such serious trouble. Endangered and threatened are two very strong words, and rare's pretty powerful, too. Sometimes I wonder why everyone isn't hysterical about the loss. Of course the reason people are so interested and involved in the stewardship network is that most times when people get an inkling of how important action is, they don't have a clue as to how to do it. The hole in the ozone, for instance: I'm upset about it, but I really don't know what I can do about it besides political things. The Volunteer Stewardship Network offers a way to act, and not just in a tiny role.

"This year we're starting a five-year project as a part of the federal recovery plan for the prairie white-fringed orchid [*Plantanthera leucophaea*], which is threatened. Illinois is mostly where it used to be—we're not on the edge of its range or anything—there should be lots of it. Volunteers will be going out to where there are a few remaining populations and pollinating the flowers, with toothpicks. We'll be coming back later and collecting a portion of those ripe seeds and taking them out to places where that plant should be

and once was and planting those seeds and monitoring the plant-
ings year after year. About five years later we may see some flowers—that's about how long it takes."

Ross later added that "one reason the plant is so rare today is
that the hawkmoths that pollinate them are not finding the widely
scattered populations. The main reason that *they're* so rare is that
their habitat has become so degraded. As we improve the habitat
through stewardship the orchids will be able to thrive. Pollination
guaranteed, seed dispersal enhanced, habitat improved—we're
interrupting the cycle that's eliminating this plant; a holistic
approach really. We're only restoring the plants to places where
they will be protected—through management and through legal
protection.

"If you had to pay scientists or even students to do the work, it
would be expensive—it's so time-consuming and labor-intensive.
Yet it's easy work to do. . . . You could, anyone could, save a species
from extinction!

"I like organizing," Ross continued. "It makes me feel that if
there's something wrong, there's something you can do about it. My
practice involves getting excited and jumping in head-first and
then figuring out, once I'm going down for the second time, what
parts of the work to keep and what parts to not keep. If I tried to
think it through in advance I probably would be a lot more timid.

"A lot of my job is matching up people with jobs when there's
all this work to be done, trying to figure out who would be good at
what and how to get them started in it so they'll be successful and
it will be good for them and good for the project. People really
appreciate being tempted by big juicy projects that they might not
think up for themselves.

"A lot of times you'll describe a project and people's first instinct
is to say, 'Well, I could do two percent of that.' I think they really
appreciate being led to understand that they could do sixty percent
of that—a hundred and ten percent. It helps them think more
about their abilities.

"The experiences get bigger and bigger—you're girdling whole trees, and you're setting fire to the woods, setting fire to the prairie with these big roaring flames—you're actually making choices about the course that the management will take.

"I don't mind it if we make some mistakes. Usually we'll notice it and stop making them. Somebody will learn something from it. You can eventually set it right. Some mistakes are pretty costly but it's a lost riskier to not do anything. We're pretty sure that if we don't do something it's going to be a lot worse."

In a practice of accentuating the positive to eliminate the negative, Ross said, "This year we're trying to manage Bluff Springs Fen [an extraordinarily rich ninety-acre mosaic of prairie, fen, and savanna ecosystems forty miles west of Chicago] as a model preserve. We want to get a lot of people in there but in a way that it won't hurt the preserve and so we're going to organize a whole series of little things. Every weekend there'll be something going on—there might be a Saturday morning bird walk with ten people. There might be a Friday night sedge identification walk. We want people in there because we want people to love the place. We also want people in there who are doing good things we like there, because there are also people doing really bad things in there—people with dirt bikes, people having messy parties, and doing some pretty destructive things. Cops are not the only answer to that, fences are not the answer to that, and laws aren't the answer to that. What we've found is that having people in there—our people in there—is a big part of the answer to that. People don't want to go where there are people who might see them at their mischief. The more people who know what a rare thing is and what their effect on it is, the more people are going to act appropriately." . . .

✼   ✼   ✼

To spotlight Laurel Ross, who for years volunteered for the North Branch prairie restorations before she became a full-fledged staff person at The Nature Conservancy's Illinois field office, and Steve Packard, whose intelligence and charisma have stimulated and animated much of the work, is to ask these two people to stand for a great many others. They would certainly demur from taking much of the credit for this remarkable program of earth healing. But it's only human to focus on individuals rather than on populations—the many hundreds of volunteers who bend their backs to the task—so I must ask the reader to pardon the synecdoche.

Packard and Ross both seem averse to ideology. This pragmatism, as a matter of fact, is very much the institutional culture of their employer, The Nature Conservancy, and of most land conservancies. You don't raise sums large enough for the purchase and protection of sizable tracts of land by denouncing the failings of private ownership and the excesses of capitalism. Quite possibly Ross and Packard feel that empowering thousands of volunteers to save species in northeastern Illinois, indeed to be encouraging the rescue of savanna ecosystems throughout their former range, is plenty to accomplish; and that the practice of restoration, inherently transforming, is enough.

Chapter 5

The Whole Horse

Wendell Berry

*For more than a third of a century, the leading agrarian voice in America
has been Kentucky farmer and writer Wendell Berry. From his small,
hilly farm in northern Kentucky, his native home and home to generations of his ancestors, Berry has commented on what he perceives as the
moral, social, and ecological decline of his country. His criticisms have
ranged widely, from abortion to nuclear power to free trade. Yet at the
center of his life and writings have always been the problems facing farms
like his own and small towns such as nearby Port Royal. As Berry has put
it, his work "has been motivated by a desire to make myself responsibly at
home both in this world and in my native and chosen place."*

*In this new essay, Berry explores the profound differences between an
economic system based on agrarian ideals and its consumptive, domi-
neering opposite, which Berry terms industrialism. Agrarianism, he tells
us, is not just an idea or a set of ideas; it is "a practice, a set of attitudes,
a loyalty, and a passion. . . . Whereas industrialism is a way of thought
based on monetary capital and technology, agrarianism is a way of
thought based on land."*

*In this tightly constructed and overflowing work, Berry synthesizes
the agrarian alternative, with particular reference to economic issues. He
chides the modern conservation movement for paying inadequate attention to the fundamental elements of the economy; for working within,
and hence unintentionally supporting, a system that is inherently
destructive of land, people, and culture. An economic transformation is
needed, he argues, and it is best begun at the local level and with attention to local matters.*

This modern mind sees only half of the horse—that half which may become a dynamo, or an automobile, or any other horsepowered machine. If this mind had much respect for the full-dimensioned, grass-eating horse, it would never have invented the engine which represents only half of him. The religious mind, on the other hand, has this respect; it wants the whole horse, and it will be satisfied with nothing less.

I should say a religious mind that requires more than a half-religion.

—Allen Tate, "Remarks on the Southern Religion," in *I'll Take My Stand*

One of the primary results—and one of the primary needs—of industrialism is the separation of people and places and products from their histories. To the extent that we participate in the industrial economy, we do not know the histories of our families or of our habitats or of our meals. This is an economy, and in fact a culture, of the one-night stand. "I had a good time," says the industrial lover, "but don't ask me my last name." Just so, the industrial eater says to the svelte industrial hog, "We'll be together at breakfast, I don't want to see you before then, and I won't care to remember you afterwards."

In this condition, we have many commodities but little satisfaction, little sense of the sufficiency of anything. The scarcity of satisfaction makes of our many commodities, in fact, an infinite series of commodities, the new commodities invariably promising greater satisfaction than the older ones. And so we can say that the industrial economy's most marketed commodity is satisfaction, and that this commodity, which is repeatedly promised, bought, and paid for, is never delivered. On the other hand, people who have much satisfaction do not need many commodities.

The persistent want of satisfaction is directly and complexly related to the dissociation of ourselves and all our goods from our and their histories. If things do not last, are not made to last, they can have no histories, and we who use these things can have no memories. We buy new stuff on the promise of satisfaction because we have forgotten the promised satisfaction for which we bought our old stuff. One of the procedures of the industrial economy is to reduce the longevity of materials. For example, wood, well made into buildings and furniture and well cared for, can last hundreds of years, but it is now routinely manufactured into products that last twenty-five years. We do not cherish the memory of shoddy and transitory objects, and so we do not remember them. That is to say that we do not invest in them the lasting respect and admiration that make for satisfaction.

The problem of our dissatisfaction with all the things we use is not correctable within the terms of the economy that produces those things. At present, it is virtually impossible for us to know the economic history or the ecological cost of the products we buy; the origins of the products are typically too distant and too scattered and the processes of trade, manufacture, transportation, and marketing too complicated. There are, moreover, too many good reasons for the industrial suppliers of these products not to want their histories to be known.

When there is no reliable accounting and therefore no competent knowledge of the economic and ecological effects of our lives, we cannot live lives that are economically and ecologically responsible. This is the problem that has frustrated, and to a considerable extent undermined, the American conservation effort from the beginning. It is ultimately futile to plead and protest and lobby in favor of public ecological responsibility while, in virtually every act of our private lives, we endorse and support an economic system that is by intention, and perhaps by necessity, ecologically irresponsible.

If the industrial economy is not correctable within or by its own terms, then obviously what is required for correction is a countervailing economic idea. And the most significant weakness of the conservation movement is its failure to produce or espouse an economic idea capable of correcting the economic idea of the industrialists. Somewhere near the heart of the conservation effort as we have known it is the romantic assumption that, if we have become alienated from nature, we can become unalienated by making nature the subject of contemplation or art, ignoring the fact that we live necessarily in and from nature—ignoring, in other words, all the economic issues that are involved. Walt Whitman could say, "I think I could turn and live with animals," as if he did not know that, in fact, we *do* live with animals, and that the terms of our relation to them are inescapably established by our economic use of their and our world. So long as we live, we are going to be living with skylarks, nightingales, daffodils, waterfowl, streams, forests, mountains, and all the other creatures that romantic poets and artists have yearned toward. And by the way we live we will determine whether or not those creatures will live.

That this nature-romanticism of the nineteenth century ignores economic facts and relationships has not prevented it from setting the agenda for modern conservation groups. This agenda has rarely included the economics of land use, without which the conservation effort becomes almost inevitably long on sentiment and short on practicality. The giveaway is that when conservationists try to be practical they are likely to defend the "sustainable use of natural resources" with the argument that this will make the industrial economy sustainable. A further giveaway is that the longer the industrial economy lasts in its present form, the further it will demonstrate its ultimate impossibility: every human in the world cannot, now or ever, own the whole catalogue of shoddy, high-energy industrial products that cannot be sustainably made or used. Moreover, the longer the industrial economy lasts, the more it will eat away the possibility of a better economy.

The conservation effort has at least brought under suspicion the general relativism of our age. Anybody who has studied with care the issues of conservation knows that our acts are being measured by a real and unyielding standard that was invented by no human. Our acts that are not in harmony with nature are inevitably and sometimes irremediably destructive. The standard exists. But having no opposing economic idea, conservationists have had great difficulty in applying the standard.

※ ※ ※

What, then, is the countervailing idea by which we might correct the industrial idea? We will not have to look hard to find it, for there is only one, and that is agrarianism. Our major difficulty (and danger) will be in attempting to deal with agrarianism as "an idea"—agrarianism is primarily a practice, a set of attitudes, a loyalty, and a passion; it is an idea only secondarily and at a remove. To use merely the handiest example: I was raised by agrarians, my bias and point of view from my earliest childhood were agrarian, and yet I never heard agrarianism defined, or even so much as named, until I was a sophomore in college. I am well aware of the danger in defining things, but if I am going to talk about agrarianism, I am going to have to define it. The definition that follows is derived both from agrarian writers, ancient and modern, and from the unliterary and sometimes illiterate agrarians who have been my teachers.

The fundamental difference between industrialism and agrarianism is this: Whereas industrialism is a way of thought based on monetary capital and technology, agrarianism is a way of thought based on land.

Agrarianism, furthermore, is a culture at the same time that it is an economy. Industrialism is an economy before it is a culture. Industrial culture is an accidental by-product of the ubiquitous effort to sell unnecessary products for more than they are worth.

An agrarian economy rises up from the fields, woods, and streams—from the complex of soils, slopes, weathers, connections, influences, and exchanges that we mean when we speak, for example, of the local community or the local watershed. The agrarian mind is therefore not regional or national, let alone global, but local. It must know on intimate terms the local plants and animals and local soils; it must know local possibilities and impossibilities, opportunities and hazards. It depends on and insists on knowing very particular local histories and biographies.

Because a mind so placed meets again and again the necessity for work to be good, the agrarian mind is less interested in abstract quantities than in particular qualities. It feels threatened and sickened when it hears people and creatures and places spoken of as labor, management, capital, and raw material. It is not at all impressed by the industrial legendry of gross national products, or of the numbers sold and dollars earned by gigantic corporations. It is interested in—and forever fascinated by—questions leading toward the accomplishment of good work: What is the best location for a particular building or fence? What is the best way to plow *this* field? What is the best course for a skid road in *this* woodland? Should *this* tree be cut or spared? What are the best breeds and types of livestock for *this* farm?—questions that cannot be answered in the abstract and that yearn not toward quantity but toward elegance. Agrarianism can never become abstract because it has to be practiced in order to exist.

And though this mind is local, almost absolutely placed, little attracted to mobility either upward or lateral, it is not provincial; it is too taken up and fascinated by its work to feel inferior to any other mind in any other place.

An agrarian economy is always a subsistence economy before it is a market economy. The center of an agrarian farm is the household. The function of the household economy is to assure that the farm family lives so far as possible from the farm. It is the subsistence part of the agrarian economy that assures its stability and its survival. A subsistence economy necessarily is highly diversified, and it characteristically has involved hunting and gathering as well as farming and gardening. These activities bind people to their local landscape by close, complex interests and economic ties. The industrial economy alienates people from the native landscape precisely by breaking these direct, practical ties and introducing distant dependencies.

Agrarian people of the present, knowing that the land must be well cared for if anything is to last, understand the need for a settled connection not just between farmers and their farms but also between urban people and their surrounding and tributary landscapes. Because the knowledge and know-how of good caretaking must be handed down to children, agrarians recognize the necessity of preserving the coherence of families and communities.

The stability, coherence, and longevity of human occupation require that the land should be divided among many owners and users. The central figure of agrarian thought has invariably been the small owner or small holder who maintains a significant measure of economic self-determination on a small acreage. The scale and independence of such holdings imply two things that agrarians see as desirable: intimate care in the use of the land, and political democracy resting upon the indispensable foundation of economic democracy.

A major characteristic of the agrarian mind is a longing for independence—that is, for an appropriate degree of personal and local self-sufficiency. Agrarians wish to earn and deserve what they have. They do not wish to live by piracy, beggary, charity, or luck.

In the written record of agrarianism there is a continually recurring affirmation of nature as the final judge, lawgiver, and pattern-maker of and for the human use of the earth. We can trace the lineage of this thought in the West through the writings of Virgil, Spenser, Shakespeare, Pope, Thomas Jefferson, and on into the work of the twentieth-century agriculturists and scientists J. Russell Smith, Liberty Hyde Bailey, Albert Howard, Wes Jackson, John Todd, and others. The idea is variously stated: We should not work until we have looked and seen where we are; we should honor Nature not only as our mother or grandmother but also as our teacher and judge; we should "let the forest judge"; we should "consult the Genius of the Place"; we should make the farming fit the farm; we should carry over into the cultivated field the diversity and coherence of the native forest or prairie. And this way of thinking is surely allied to that of the medieval scholars and architects who saw the building of a cathedral as a symbol or analogue of the creation of the world. The agrarian mind is, at bottom, a religious mind. It subscribes to Allen Tate's doctrine of "the whole horse." It prefers the Creation itself to the powers and quantities to which it can be reduced. And this is a mind completely different from that which sees creatures as machines, minds as computers, soil fertility as chemistry, or agrarianism as an idea. John Haines says that "the eternal task of the artist and the poet, the historian and the scholar . . . is to find the means to reconcile what are two separate and yet inseparable histories, Nature and Culture. To the extent that we can do this, the 'world' makes sense to us and can be lived in." I would add only that this applies also to the farmer, the forester, the scientist, and others.

The agrarian mind begins with the love of fields and ramifies in good farming, good cooking, good eating, and gratitude to God. Exactly analogous to the agrarian mind is the sylvan mind that begins with the love of forests and ramifies in good forestry, good woodworking, good carpentry, and gratitude to God. These two kinds of mind readily intersect and communicate; neither ever intersects or communicates with the industrial-economic mind. The industrial-economic mind begins with ingratitude and ramifies in the destruction of farms and forests. The "lowly" and "menial" arts of farm and forest are mostly taken for granted or ignored by the culture of the "fine arts" and by "spiritual" religions; they are taken for granted or ignored or held in contempt by the powers of the industrial economy. But in fact they are inescapably the foundation of human life and culture, and their adepts are capable of as deep satisfactions and as high attainments as anybody else.

Having, so to speak, laid industrialism and agrarianism side by side, implying a preference for the latter, I will be confronted by two questions that I had better go ahead and answer.

The first is whether or not agrarianism is simply a "phase" that we humans had to go through and then leave behind in order to get onto the track of technological progress toward even greater happiness. The answer is that although industrialism has certainly conquered agrarianism, and has very nearly destroyed it altogether, it is also true that in every one of its uses of the natural world industrialism is in the process of catastrophic failure. Industry is now desperately shifting—by means of genetic engineering, global colonialism, and other contrivances—to prolong its control of our farms and forests, but the failure nonetheless continues. It is not possible to argue sanely in favor of soil erosion, water pollution, genetic impoverishment, and the destruction of rural communities and local economies. Industrialism, unchecked by the affections and concerns of agrarianism, becomes monstrous. And this is because of a weakness identified by the Twelve Southerners of *I'll Take My Stand* in their "Statement of Principles": Under the rule of industrialism "the remedies proposed . . . are always homeopathic." Industrialism always proposes to correct its errors and excesses by more industrialization.

The second question is whether or not by espousing the revival of agrarianism we will commit the famous sin of "turning back the clock." The answer to that, for present-day North Americans, is fairly simple. The overriding impulse of agrarianism is toward local adaptation of economies and cultures. Agrarian people wish to fit the farming to the farm and the forestry to the forest. At times and in places we latter-day Americans may have come close to accomplishing this goal, and we have a few surviving examples, but it is generally true that we are much further from local adaptation now than we were fifty years ago. We never yet have developed stable, sustainable, locally adapted land-based economies. The good rural enterprises and communities that we find in our past have been almost constantly under threat from the colonialism, first

foreign and then domestic and now "global," that has so far dominated our history and that has been institutionalized for a long time
in the industrial economy. The possibility of an authentically settled
country still lies ahead of us.

❖   ❖   ❖

If we wish to look ahead, we will see not only in the United States
but in the world two economic programs that conform pretty
exactly to the aims of industrialism and agrarianism as I have
described them.

The first is the effort to globalize the industrial economy, not
merely by the expansionist programs of supranational corporations
within themselves but also by means of government-sponsored
international trade agreements, the most prominent of which is the
World Trade Organization Agreement, which institutionalizes the
industrial ambition to use, sell, or destroy every acre and every creature of the world.

The World Trade Organization gives the lie to the industrialist
conservatives' professed abhorrence of big government. The cause
of big government, after all, is big business. The power to do large-
scale damage, which is gladly assumed by every large-scale industrial enterprise, calls naturally and logically for government regulation, which of course the corporations object to. But we have a good
deal of evidence also that the leaders of big business actively desire
and promote big government. They and their political allies, while
ostensibly working to "downsize" government, continue to promote government helps and "incentives" to large corporations and,
however absurdly, to adhere to their notion that a small government, taxing only the working people, can maintain a big highway
system, a big military establishment, a big space program, and
award big government contracts.

But the most damaging evidence is the World Trade Organization itself, which is in effect a global government with power to

enforce the decisions of the collective against national laws that conflict with it. The coming of the World Trade Organization was foretold seventy years ago in the "Statement of Principles" of *I'll Take My Stand,* which said that "the true Sovietists or Communists . . . are the industrialists themselves. They would have the government set up an economic super-organization, which in turn would become the government." The agrarians of *I'll Take My Stand* did not foresee this because they were fortune-tellers but because they had perceived accurately the character and motive of the industrial economy.

The second program, counter to the first, is composed of many small efforts to preserve or improve or establish local economies. These efforts on the part of nonindustrial or agrarian conservatives, local patriots, are taking place in countries both affluent and poor all over the world.

Whereas the corporate sponsors of the World Trade Organization, in order to promote their ambitions, have required only the hazy glamour of such terms as *the global economy, the global context,* and *globalization*—and thus apparently have vacuum-packed the minds of every politician and political underling in the world—the local economists use a much more diverse and particularizing vocabulary that you can actually think with: *community, ecosystem, watershed, place, homeland, family, household.*

And whereas the global economists advocate a world-government-by-economic-bureaucracy, which would destroy local adaptation everywhere by ignoring the uniqueness of every place, the local economists found their work upon respect for such uniqueness. Places differ from one another, the local economists say, and therefore we must behave with unique consideration in each one; the ability to tender an appropriate practical regard and respect to each place in its difference is a kind of freedom; the inability to do so is a kind of tyranny. The global economists are the great centralizers of our time. The local economists, who have so far attracted the support of no prominent politician, are the true decentralizers and downsizers, for they seek an appropriate degree of self-determination and independence for localities. They seem to be moving toward a radical and necessary revision of our idea of a city. They are learning to see the city not as a built and paved municipality set apart by "city limits" to live by trade and transportation from the world at large, but rather as a part of a community that includes also the city's rural neighbors, its surrounding landscape and its watershed, on which it might depend for at least some of its necessities, and for the health of which it might exercise a competent concern and responsibility.

At this point, I want to say point-blank what I hope is already clear: Although agrarianism proposes that everybody has agrarian responsibilities, it does not propose that everybody should be a farmer or that we do not need cities. Nor does it propose that every product be a necessity. Furthermore, any thinkable human economy would have to grant to manufacturing an appropriate and honorable place. Agrarians would insist only that any manufacturing enterprise should be formed and scaled to fit the local landscape, the local ecosystem, and the local community and that it should be locally owned and employ local people. They would insist, in other words, that the shop or factory owner should not be an outsider but rather a sharer in the fate of the place and its community. The deciders should live with the results of their decisions.

Between these two programs—the industrial and the agrarian, the global and the local—the most critical difference is that of knowledge. The global economy institutionalizes a global ignorance, in which producers and consumers cannot know or care about one another, and in which the histories of all products will be lost. In such a circumstance, the degradation of products and places, producers and consumers, is inevitable.

But in a sound local economy, in which producers and consumers are neighbors, nature will become the standard of work and production. Consumers who understand their economy will not tolerate the destruction of the local soil or ecosystem or watershed as a cost of production. Only a healthy local economy can keep nature and work together in the consciousness of the community. Only such a community can restore history to economics.

❋ ❋ ❋

I will not be altogether surprised to be told that I have set forth here a line of thought that is attractive but hopeless. A number of critics have advised me of this, out of their charity, as if I might have written of my hopes for forty years without giving a thought to hopelessness. Hope, of course, is always accompanied by the fear of hopelessness, which is a legitimate fear.

And so I would like to conclude by confronting directly the issue of hope. My hope is most seriously challenged by the fact of decline, of loss. The things I have tried to defend are less numerous and worse off now than when I started, but in this I am only like all other conservationists. All of us have been fighting a battle that on average we are losing, and I doubt that there is any use in reviewing the statistical proofs. The point—the only interesting point—is that we have not quit. Ours is not a fight that you can stay in very long if you look on victory as a sign of triumph or on loss as a sign of defeat. We have not quit because we are not hopeless.

My own aim is not hopelessness. I am not looking for reasons to give up. I am looking for reasons to keep on. In outlining here the concerns of agrarianism, I have intended to show how the effort of conservation could be enlarged and strengthened.

What agrarian principles implicitly propose—and what I explicitly propose in advocating those principles at this time—is a revolt of local small producers and local consumers against the global industrialism of the corporations. Do I think there is hope that such a revolt can survive and succeed, and that it can have a significant influence upon our lives and our world?

Yes, I do. And to be as plain as possible, let me just say what I know. I know from friends and neighbors and from my own family that it is now possible for farmers to sell at a premium to local customers such products as "organic" vegetables, "organic" beef and lamb, and pasture-raised chickens. This market is being made by the exceptional goodness and freshness of the food, by the wish of urban consumers to support their farming neighbors, and by the excesses and abuses of the corporate food industry.

This, I think, gives the pattern of an economic revolt that not only is possible but is happening. It is happening for two reasons: First, as the scale of industrial agriculture increases, so does the scale of its abuses, and it is hard to hide large-scale abuses from consumers. It is virtually impossible now for intelligent consumers to be ignorant of the heartlessness and nastiness of animal confinement operations and their excessive use of antibiotics, of the use of hormones in meat and milk production, of the stenches and pollutants of pig and poultry factories, of the use of toxic chemicals and the waste of soil and soil health in industrial row-cropping, of the mysterious or disturbing or threatening practices associated with industrial food storage, preservation, and processing. Second, as the food industries focus more and more on gigantic global opportunities, they cannot help but overlook small local opportunities, as is made plain by the increase in community-supported agriculture, farmers' markets, health food stores, and so on. In fact, there are some markets that the great corporations by definition cannot supply. The market for so-called organic food, for example, is really a market for good, fresh, trustworthy food, food from producers known and trusted by consumers, and such food cannot be produced by a global corporation.

But the food economy is only one example. It is also possible to think of good local forest economies. And in the face of much neglect, it is possible to think of local small business economies—some of them related to the local economies of farm and forest—supported by locally owned, community-oriented banks.

What do these struggling, sometimes failing, sometimes hardly realized efforts of local economy have to do with conservation as we know it? The answer, probably, is *everything*. The conservation movement, as I said earlier, has a conservation program; it has a preservation program; it has a rather sporadic health-protection program; but it has no economic program, and because it has no economic program it has the status of something exterior to daily life, surviving by emergency, like an ambulance service. In saying this, I do not mean to belittle the importance of protest, litigation, lobbying, legislation, large-scale organization—all of which I believe in and support. I am saying simply that we must do more. We must confront—on the ground, and each of us at home—the economic assumptions in which the problems of conservation originate.

We have got to remember that the great destructiveness of the industrial age comes from a division, a sort of divorce, in our economy, and therefore in our consciousness, between production and consumption. Of this radical division of functions we can say, without much fear of oversimplifying, that the aim of producers is to sell as much as possible and that the aim of consumers is to buy as much as possible. We need only add that the aim of both producer and consumer is to be so far as possible carefree. Because of various pressures, governments have learned to coerce from producers some grudging concern for the health and solvency of consumers. No way has been found to coerce from consumers any consideration for the methods and sources of production.

What alerts consumers to the outrages of producers is typically some kind of loss or threat of loss. We see that in dividing consumption from production we have lost the function of conserving. Conserving is no longer an integral part of the economy of the producer or that of the consumer. Neither the producer nor the consumer any longer says, "I must be careful of this so that it will last." The working assumption of both is that where there is some, there must be more. If they can't get what they need in one place, they will find it in another. That is why conservation is now a separate concern, a separate effort.

But experience seems increasingly to be driving us out of the categories of producer and consumer and into the categories of citizen, family member, and community member, in all of which we have an inescapable interest in making things last. And here is where I think the conservation movement (I mean that movement that has defined itself as the defender of wilderness and the natural world) can involve itself in the fundamental issues of economy and land use, and in the process gain strength for its original causes.

I would like my fellow conservationists to notice how many people and organizations are now working to save something of value—not just wilderness places, wild rivers, wildlife habitat, species diversity, water quality, and air quality but also agricultural land, family farms and ranches, communities, children and childhood, local schools, local economies, local food markets, livestock breeds and domestic plant varieties, fine old buildings, scenic roads, and so on. I would like my fellow conservationists to understand also that there is hardly a small farm or ranch or locally owned restaurant or store or shop or business anywhere that is not struggling to save itself.

All of these people, who are fighting sometimes lonely battles to preserve things of value that they cannot bear to lose, are the conservation movement's natural allies. Most of them have the same enemies as the conservation movement. There is no necessary conflict among them. Thinking of them, in their great variety, in the essential likeness of their motives and concerns, one thinks of the possibility of a defined community of interest among them all, a shared stewardship of all the diversity of good things that are needed for the health and abundance of the world.

I don't suppose that this will be easy, given especially the history of conflict between conservationists and land users. I suppose only that it is necessary. Conservationists can't conserve everything that needs conserving without joining the effort to use well the agricultural lands, the forests, and the waters that we must use. To enlarge the areas protected from use without at the same time enlarging the areas of *good* use is a mistake. To have no large areas of protected old-growth forest would be folly, as most of us would agree. But it is also folly to have come this far in our history without a single working model of a thoroughly diversified and integrated, ecologically sound, local forest economy. That such an economy is possible is indicated by many imperfect or incomplete examples, but we need desperately to put the pieces together in one place—and then in every place.

The most tragic conflict in the history of conservation is that between the conservationists and the farmers and ranchers. It is tragic because it is unnecessary. There is no irresolvable conflict here, but the conflict that exists can be resolved only on the basis of a common understanding of good practice. Here again we need to foster and study working models: farms and ranches that are knowledgeably striving to bring economic practice into line with ecological reality, and local food economies in which consumers conscientiously support the best land stewardship.

We know better than to expect very soon a working model of a conserving global corporation. But we must begin to expect—and we must, as conservationists, begin working for, and in—working models of conserving local economies. These are possible now. Good and able people are working hard to develop them now. They need the full support of the conservation movement now. Conservationists need to go to these people, ask what they can do to help, and then help. A little later, having helped, they can in turn ask for help.

Chapter 6

What Comes Around

Gene Logsdon

The rolling farmlands of Ohio have produced over the decades a steady flow of literate agrarian spokesmen. During the middle decades of the twentieth century, native Louis Bromfield, a Pulitzer Prize–winning novelist, ably carried the agrarian banner from his expansive Ohio farm, Malabar, where he wrote lyrically of the joys of rural life and the indispensable need to farm in ways that protect and build fertile soil. Today, Gene Logsdon is among those carrying forward this valuable tradition. The self-styled "Contrary Farmer," Logsdon roams far in his critique of modern culture even as he offers particularized, place-specific advice on organic gardening and other homestead skills.

In the following piece, Logsdon speaks of the need to know the history of one's chosen place and to embrace a long-term perspective toward it. The story of Logsdon's home ground has been one of change, particularly in patterns of landownership and land use. The primary lesson Logsdon draws is that today's industrial agriculture and large-scale landholdings are neither inevitable nor likely to endure. Large scale can give way to small scale, just as it has done in the past; high tech can give way to low tech; corporate farming can yield to the re-emergence of family enterprises. What comes around can go around.

Logsdon looks forward to a new generation of "countrysiders" who combine part-time, family-centered food production with other employment, giving shape to modes of living that "join the best of urban life with the best of rural life in a new and admirable agrarianism."

Thoughts while contemplating a woodlot that forty-four
years ago was the highest-yielding cornfield in the world.

Three years ago, a nationally renowned agricultural economist made a prediction on the radio that I have a hunch will embarrass him greatly if he lives long enough. (Perhaps all our attempts at predicting the future will embarrass us greatly if we live long enough.) He said that a continuation of larger and larger industrial grain farms and animal factories was "inevitable." It was obvious that inevitable also meant irreversible in his mind. He did not make this statement as his opinion, but as a fact, one that sentimental old agrarian cranks like me had better get used to.

I wonder if he would have made that prediction had he known the history of any one farm deeply or if he had known that the proprietors of the factory farms he was so arrogantly extolling in 1997 would be standing like bums in a soup line in 1998, waiting for huge government handouts to keep them financially afloat. In 1999, almost half the income farmers received came not from farming, but from government.

"Inevitable" is a word that probably ought to be stricken from the language of human behavior. History demonstrates, time and time again, that in agriculture, as in any economic activity, change is the only inevitability. It is just as possible for farming to go from big to small in size as from small to big—to disperse into many units as to consolidate into a few. Nor is the supposed normal progression of land development from wildland to metropolis inevitable and irreversible. I need look no farther than right here in the fields of home to see that lesson written on the land. If Wyandot Chief War Pole, after whom the creek that runs through our farm is named, had been a conventional economist, he would have insisted, fifty years before the influx of white settlement, that the continuation of the Wyandots' highly refined combination of farming and hunting was inevitable and the Delawares and Shawnees had better get used to it. But War Pole lived to see sheep ranching become the characteristic agricultural activity here after the Wyandots were tragically shipped off to Kansas in the mid-1800s. War Pole's people and their ancestors had unwittingly prepared for the ranchers by creating, with their annual fire-ring hunts, extensive treeless prairie pastures ideal for sheep. My great-grandfather Charles Rall, coming here from Germany, leading two cows all the way from Columbus, went to work on the R.M. Taylor ranch that spread over much of the farmland I have roamed since childhood. Had there been agricultural economists in those days, I can just hear them saying, with all the pomp and ceremony of their royal offices, that a continuation of huge sheep ranches was our "inevitable" future and hired hands from Germany better get used to it.

But within a generation, money was finding different paths to follow. The sheep ranches rapidly became "obsolete" (in human civilization, obsolete means "something no longer money-profitable enough to compete with something else"), and farmers like my great-grandfather bought up the prairie along with the forests around it and converted it to more-diversified, large livestock farms. So profitable was this kind of farming for a while that by 1900, Great-grandfather had consolidated some 2,000 acres into his operation, and his was not the largest. Get big or get out, I can hear the economists boasting as if they were responsible: a continuation of large grain and livestock farms was "inevitable." But Great-grandfather's four sons inherited that land, divided it into as many parcels, and eventually distributed it out to their numerous offspring in 160-acre family farms, as I described earlier. That was the most "efficient" way to apply humanpower for profitability at that time. By then, agricultural economists were on the scene, having found a way to milk the tax rolls for their salaries, and they rallied to that redistribution of land with gusto, declaring its continuation not only inevitable, but a great victory for American free enterprise.

Not quite two generations later, after World War II, "free enterprise" began displacing these little farms with industrial cash grain operations headed back in size to about the acreage of Great-grandfather's farm. The economists, once more displaying not a whit of historical sense, said, and say, that the continuation of this megafactory farming is "inevitable."

I can gaze up Warpole Creek from my farm and see the once-forested valley that livestock farming kept in pasture for one hundred years. In that pasture rises a Hopewell Indian mound whose people would have been clearing this land and planting corn five hundred years before it grew back to forest and a thousand years before Great-grandfather (or perhaps the Wyandots) cleared it again. When the sun slants low in the west, I can see Great-uncle Albert's old dead furrows under the grass of the valley slopes. He tried to grow corn on these hillsides too, until he understood such land was better kept in pasture. He even tried to grow corn on the mound itself! Today this little valley, found not to be accessible enough for the monster machines of industrial grain farming (after a brief flirtation in that direction), is growing back to timber again!

Whose agrarian vision do you want to vote for? Reading what the highly literate Wyandots said about their way of life, I am convinced that they not only had effected the most ecological farming ever done here but also were the happiest farmers. Of course, the Hopewells' agrarianism might have been just as pleasant as the Wyandots', for a while. There is no way to know, because they did not leave a written record of their marvelous integration of farming and commerce and hunting-gathering, an economy that archaeologists say was so successful that it eventually (inevitably?) generated overpopulation and collapsed. But you can bet that right up until their decline began, maybe even after it started, the shamans who passed for economists in the Hopewell villages stood atop their mounds and declared with immense bravado that a continuation of bigger and more elaborate earthworks was "inevitable."

✳   ✳   ✳

Between the shift from Great-grandfather's large livestock farms to his grandchildren's small family farms, something else happened that further reveals the sham of economic prediction that ignores the history of place. The corn and soybean fields that I can look out upon from my eastward windows today were for a time an airfield! It was known as "Rall Field," naturally enough. The year was 1930. The prophets of inevitability were all talking about how there would be a plane as well as a car in every garage someday.

My kinfolks' airport is remembered not because it violated the conventional theories of historical progression, but because of a humorous story that went with it. On Sundays, planes would fly in from Bucyrus and Marion and other towns in the area and take people for rides. The planes were mostly fragile, homemade affairs guaranteed to supply plenty of weekend excitement. The story goes that the owner-builder of one such plane, possibly not trusting the flimsy crate himself, hired a pilot to fly it from Bucyrus, where he kept it, to Rall Field for an afternoon of rides. Arriving at the field before his plane, the budding airline executive noticed a dead furrow across the upper end of the landing strip. Though only a slight depression in the ground, it might spell disaster if the plane crossed over it during landing. So the first and only air traffic controller Mifflin Township has ever known straddled the worrisome little remnant of bygone agrarianism, and as the plane hove into sight began waving his hands and pointing down to the ground at the source of his consternation. The pilot interpreted the pantomime in just the opposite way it was intended. It seemed strange to him that his boss wanted the plane set down right in front of what looked like a dead furrow, but it was obvious, from the increasing ferocity with which he waved, that such was the case. Down he came, as close as he could to the spot his screaming, purple-faced air traffic controller was pointing at. When the plane hit the furrow, it nosed over and crumpled up like a paper accordion, but the rate of speed was so slow that the pilot walked away unhurt.

Rall Field did not last nearly as long as the vision of an inevitable plane in every garage. Great-uncle Albert, applying his astute pencil stub to the daybook he kept handily in his bib overalls, calculated that corn and dairy cows on that land were more profitable than inevitable airplanes, at least for the time being. Today, he might have concluded that a golf course was more inevitable than dairy cows, except that other farmers not far away have already reached that conclusion. There are now more golf courses in our county per resident (24,000 population, four golf courses) than possibly anywhere in the nation. This was one "inevitability" entirely missed by the economic prognosticators.

❧ ❧ ❧

Many farms have strange tales to tell. The first officially recorded 300-bushel corn yield was grown by Lamar Ratcliff on his father's farm in Prentiss County, Mississippi (much to the chagrin of the cornbelt), back in 1955. How well I remember the excitement among us farmers. The farm magazine rhetoric flowed with the promise of mighty things to come. Soon 300-bushel corn would be common. And if Mississippi could do it, by hickory Illinois and Iowa with the help of more fertilizers and chemicals and hybrid vigor and technology, yawn yawn, would soon ring up 400-bushel yields. The word "inevitable" was flung around very loosely on that occasion too.

Today, forty-four years later, the field that grew the first 300-bushel corn is a woodlot again! Furthermore, yields of 300 bushels per acre were not achieved again for twenty years and then only in a dozen or so isolated instances. Ironically, agronomic experimentation indicates that if 300-bushel yields ever do become "inevitable," they will be a product of biointensively managed, raised-bed garden plots, where extremely high yields of almost everything have been achieved in recent years. Imagine how much food the backyards of America alone could produce at such super yields (or even at average yields), putting the lie to the claim by the mega-food companies and their university hirelings that big business is the only and inevitable solution to the future of food production.

We have been here before and know this lie. Way back in 1907, an economist in England, Prince Kropotkin, a forerunner of the new breed of ecological economists today, clearly demonstrated with pages of data how food and manufactured goods could be produced just as abundantly and economically on small, dispersed farms and shops as on the huge bonanza acreages and large inhumane factories that greed was generating at that time. His book, *Fields, Factories, and Workshops,* correctly predicted the demise of the bonanza farms and the "farming out" by large factories of much of their manufacturing work to smaller, more efficient shops. Consolidation is not a synonym for efficiency but only for power.

That the current trend to consolidation in the food business will "inevitably" reverse itself is as justifiable a conclusion as assuming that a half dozen huge companies will "inevitably" monopolize and control the food supply. All over the "inevitably" industrialized cash grain county I wander daily, small homesteads and garden farms are popping up. Some of them are merely urban homes in the country, but a surprising number are younger couples or retirees coming back to reclaim and use some of the land that the economics of power took away from their fathers and grandfathers. When farms go up for sale around here now, they are invariably split up into small parcels of five to forty acres, because there is such a demand for such acreage. Young families with urban jobs can and will bid more for the small parcels than the mega-farmers can afford to pay, a reversal of a thirty-year trend. The mega-farmers fume at the practice of selling farms in small parcels, forgetting how they said, in their day of breezily buying every farm that came up for sale, that this is just "the good ole' American way."

Why this new development in rural life receives little attention from the economists is beyond me. (Not really. The people spearheading this "forward to the land" movement do not want to ask for political or educational help and so of course the bureaucrats must pretend they don't exist to save face.) My own family makes a good example. My parents raised nine children on a typical family farm of the thirties and forties. Eight of us children live rural lives today much like our parents did, two of us heavily into commercial farming but with other sources of income, and the rest of us dividing our time between other careers and small-scale farming, sheep raising, orcharding, tree farming, and very serious, subsistence gardening. Of our twenty-nine children, fourteen are married, and of that number, nine live rural lives as we do, and three of the five others tell me they will move forward to the land as soon as they save sufficient money to do so. Of the remaining fifteen still in school or still single and trying to find their way, I am certain that at least half will eventually (inevitably?) take up our rural lifestyle. And we are outdone in this respect by other rural-rooted families in our neighborhood.

We "new" countrysiders express an allegiance to the same agrarian values that our parents and grandparents and great-grandparents honored. The only difference is in our way of expressing that allegiance and in the work we pursue to achieve it. We come back to rural life because we want some physical control over our lives. We are rebelling against the economics of power. We want some income from the land but also some from nonfarm sources because we understand the folly of trying to make a decent financial income entirely from farming in today's power economy. We want homes where our children can know meaningful work and learn something useful as they grow up. We want an alternative to chemicalized, hormonized, vaccinized, antibiotic-treated, irradiated factory food. We would like to establish home-based businesses when possible so that we do not have to put our children in day care centers as parents who work away from home must often do. We want a different kind of educational environment for our children from what consolidated, power-economics schools provide, private or public. Sometimes we homeschool our children. We want, above all, some home-based security not dependent entirely on power economics. We think the economics of power has run its course in this cycle and is going to hell.

What we are doing, in short, is finding ways, which the early farm organizations failed to do because they taught farmers to put all their eggs in one economic basket, to bring back to rural America the life and money that consolidated banking sucked out over two centuries of predatory colonization, a progress that consolidated schooling legitimized. We are managing to join the best of urban life with the best of rural life in a new and admirable agrarianism. Steve Zender, editor of one of our local village newspapers, wrote a telltale anecdote in a book he published in 1998 in celebration of rural and village life, *The Big Things in Life Are the Little Things.* One of his reporters, Kate Orians, left a message on his office phone: "I'll bring my story over Sunday evening or afternoon. It just got too late today and the vet was here to take blood samples from the pigs." Zender laughs as he tells the story again. "It is wonderful to live in rural America where we have the best of both worlds," he says. "Futurists say that telecommunications will result in people fleeing the cities and moving to rural areas. It's already happening here and I hope it means that people will be able to provide for their families and do important work for their communities while still having time to garden and take care of pigs."

The most interesting and promising ideas in food production are showing up in this "new" agrarian society. While mainstream factory farming continues to cement itself financially into huge cumbersome operations that lose the flexibility to move quickly to take advantage of changes in consumer-driven markets, small-scale farms are perfecting new/old practices like deep-bedding systems for hogs that are free of factory farm odor and pollution problems; organic dairies; meat, milk, and egg production that relies on rotational pasture systems, not expensive chemicals or machinery; permanent, raised-bed vegetable gardens where production per square foot is enormously increased with hand labor, not expensive machinery; improved food plants from natural selection of open-pollinated varieties whose seed can be economically saved for the next year's crop; and so forth. The mind-set of the new small farmers is not simply traditional. The seminar that drew the biggest crowd at a recent small farm conference in Indiana was on raising freshwater prawns. The thinking of these new farmers is rather untraditional, too. Chip Planck, a successful, long-standing, commercial market gardener near Washington, D.C., used to be a professor of political science.

Even the schools of economics are headed in a different direction from what many of their conventional economists realize. New economic theories have reverence for historical evidence. One of the foremost spokespersons of this different philosophy of economics, John Ikerd at the University of Missouri, recently wrote a paper titled "The Coming Renaissance of Rural America." Discussing change in food production cultures, he asserted the validity of the "universal cycle theory," which *Science* magazine recently included in its list of the top twenty scientific ideas of the twentieth century. According to this theory, any observed trend is in fact just a phase of a cycle. "If we look back over the past centuries and around the globe, we can find examples where control of land became concentrated in the hands of a few only to later become dispersed in control among the many," writes Ikerd. "The trend toward fewer and larger farms in the U.S. might be just a phase of a cycle that is nearing its end."

The ending of this cycle, as the ending of any cycle, will be sad, even tragic, for those who do not see it coming and so do not change in time. The gigantic, high-investment food factories of today, like the 10, 20, 30, and even 75 thousand-acre bonanza farms at the turn of the twentieth century, do not in fact have the flexibility *now* to respond fast enough to changes in market demand. If just 25 percent of the people in this country decided to become vegetarians, or if 40 percent decided to cut their consumption of meat significantly, which is entirely likely, the factory farm system irrevocably based on meat, corn, and soybean meal is history. Like bonanza farms, like any dinosaur without great adaptability to a new environment, such a system must die out until conditions are right for it again.

But in every other way than the financial welfare of the bonanza crowd, a swing back to a more distributive food production system will have many advantages for the common good, not the least of which would be an end (until the next cycle) of the food monopolies now being created.

The problem I fear is that while we are condemned by the economics of money greed, or changing markets, or weather patterns, to continue the inexorable cycles of small and large, boom and bust, consolidation and dispersal, the power of wealth solidified in the current cycle will try to stop the wheel of history from turning to the next cycle. Expect a vigorous effort on the part of the agribusiness oligarchy, in cahoots with a suppliant government, to prolong the bonanza farm consolidation of today as long as it can. Some of this futility will be justified to prevent a period of chaos in readjustment, just as the futile attempt to "save the family farm" of the last cycle was justified for a similar reason. But prolonging the end of the current cycle of consolidated power will be much more potentially dangerous because those being "helped" will not be family farmers but the wealthy classes who need no help. Expect the power structure to continue the present policy of giving welfare capital shamelessly to the rich in the name of saving society from starvation. Expect it to continue to perpetrate propaganda that hides the collapsing bonanza economy, especially the health hazards of factory food. Expect it to continue legal favoritism to the food monopolies.

In any event, there is one consolation for old agrarian cranks like me. Even under continued factory farm economics, monopoly can't work very well in food production as long as enough people have access to land, even backyard land, to grow food for themselves and local customers. We may not be able to make cars in our backyards cheaper than the moguls of money can do it, but give us land, any land, even the industrially bombed-out rubble of inner cities, and we can compete with factory food. Hardly anything is simpler or cheaper to do than raising chickens and vegetables.

Chapter 7

The Urban-Agrarian Mind

David W. Orr

The agrarian concern for the land's health calls into question urban land-use practices as well as rural ones. Resources flow into cities and wastes flow out, affecting distant lands in ways invisible to most city dwellers. Professor and environmental visionary David Orr is among those working to make cities and towns—and hence the planet as a whole—more livable by heightening awareness of adverse effects on distant lands and searching for ways to reduce them.

This new essay was originally delivered as a talk at The Land Institute (see chapter 1) in 1999. Orr describes how agrarian goals and modes of thought can be promoted by drawing selectively on the most advanced tools of industrial culture, using as his chief illustration the energy-efficient, land-sensitive environmental studies center recently constructed under his guidance at Oberlin College. He also considers the daunting challenge of instilling an awareness of land issues in college students whose lives have been shaped by malls, highways, and computers. Like Gene Logsdon, Orr believes that land-use and design mistakes of the past need not go uncorrected; as he urges here, quoting an inner-city youth, "asphalt isn't forever."

Among the more encouraging cultural trends observable today is a gradual resurgence of agrarian values and practices. Agrarianism as a state of mind, as a cultural perspective, is rooted in land and respectful in fundamental ways of the land's possibilities, mysteries, and limits. It has long been opposed and indeed, as we know, overpowered by a clever, market-driven industrial mentality that perceives no natural limits and treats land as mere raw material. I am encouraged by this revival of agrarianism, however modest— encouraged both because its fundamental values are right and because it builds on perhaps the only secular tradition in American culture that speaks for land and its long-term health. Yet mixed with this encouragement are some sober recognitions: of the strength of the industrial mentality and of what it will take for agrarian values to regain real influence.

The agrarian mind arose some ten thousand years ago, chiefly out of necessity. It was the product, so far as we know, neither of intention nor of formal learning. But if agrarian values and prac- tices are to play a meaningful role in our future, the agrarian mind will need to be rediscovered, dusted off, and adapted to new cir- cumstances. Led by places such as The Land Institute, it will need to be deliberately reconstructed and applied in new settings while remaining true to its core values and insights. This rediscovery of agrarianism will not be a matter merely of reform or tinkering at the margins. It calls for a fundamental re-visioning of how we per- ceive our place in nature and how we provision ourselves with food, energy, and materials.

Let me describe our present situation by relating a story told by environmental observer Peter Montague. Our situation, Montague says, is this: We're all passengers on a long, rickety train going south at forty miles per hour, not rushing toward doom but steadily chug- ging southward toward environmental and social destruction. Many of us are alert to the dangers, and for several years we've been earnestly walking north inside the train. As we plod from train car to train car, we stop to congratulate ourselves on our progress, slap- ping one another on the back or hugging as we recount the many train cars we've managed to pass through. But if we would only pause to look out the window, we would see that we're now farther south than we were when we last stopped to congratulate ourselves. Despite our best efforts, we've been unable to reverse direction.

Maybe this is happening because we've spent our time engaging the conductor in conversation. This seems like a natural thing to do; after all, it's the conductor who sets and enforces the rules on the train. Furthermore, the conductor seems intelligent and genuinely interested in helping us make our way north through the train. He emphasizes how well we are doing, and when we become discouraged he urges us on, reminding us that walking northward is a noble venture and hinting that in time, we'll reach the promised place. Unfortunately, it has been years since we asked ourselves the fundamental questions: What fuels the locomotive? Who is the engineer with his hand on the throttle? Why is he still leading us southward when we know the direction is wrong? And what will it take to make him change direction?

After several decades of Prairie Festivals and similar gatherings elsewhere, we are still on that southbound train. We have not yet really come to know the conductor, let alone formed a strategy to reverse direction. What we have learned, I think, is that turning the train around will require a change in our industrial, market-based mentality, more fundamental in nature than any change we've imagined so far. The best way to understand that change is to conceive of it as a marriage of our dominant industrial mind-set with a far older agrarian mind. The practical questions are whether, how, and on what terms that union might take place.

To help us think about such a marriage, as we stand here at the courtship stage, let me offer six observations.

First, those calling themselves agrarians, from Hesiod to Wendell Berry, all agree that we know of no stable or decent way to organize human affairs that is not somehow rooted in a deep, practical respect for soil.

Second, urbanism and industrialism—and consequently the urban-industrial mind—are now triumphant virtually everywhere.

Third, that victory has impoverished senses of place, practical skills, and indeed entire categories of thought rooted in ecological competence.

Fourth, this triumph is commonly regarded as moral progress because, among other things, it helped eliminate slavery: Yankee factories brought down the plantation-based South. Other good things were associated with industrialization, and they should not be minimized. But industrialization gave rise to moral ills hardly less destructive than slavery. Powered by fossil fuels, the industrial world has spun off biotic impoverishment, degraded landscapes, depleted resources, and a climate altered to the great disadvantage of our descendants. Although not exactly slavery, these conditions are nonetheless a kind of bondage, with the notable and aggravating difference that in time they'll burden everyone on earth, with no possibility of full release.

Fifth, the urban-industrial world is in the process of gradual failure, a failure that will become catastrophic if allowed to run its course. What we don't know is how high we'll let the flood tides go and what will happen after the tides recede.

Finally, we cannot simply turn the train around by going back to an older way of life termed agrarian. Agrarian traditions, skills, and memories have largely been forgotten. In any case, agrarian life was never as good as we sometimes imagine. Allen Tate, I believe, had it right: A true agrarian world is yet to be created. To do that, we'd need to understand, for starters, why the family farm failed. For agricultural researcher Marty Strange, the causes of that failure were rooted in misguided farm policies and unfavorable economics. He is right, no doubt, but other forces were also at work. In a capitalist economy, life on the land was often hard, insecure, and lonely. It was particularly hard on children and women. And it was rooted in the displacement of native peoples, and hence on injustice. Were we to probe more deeply, we might find that farming has always been tenuous; it was a recent evolutionary adaptation, poorly calibrated with the hunter-gatherer, tribal natures bred into us over 99 percent of our evolutionary career. It is possible, as Wes Jackson and Paul Shepard have argued, that farming as we now do it just doesn't suit our paleolithic dispositions. Whatever the cause, a durable land-based culture never took shape in the United States.

And the good parts that did take shape now live on more as memories than as actual practices. The reality is that we start mostly from ruins.

What might a new and better agrarianism be like? For one thing, it is going to be less male dominated; it will involve new gender relationships, which will be good. As the fossil-fuel era lurches to its finale, farms are likely to become smaller. For reasons built into the energetics of soil and sunlight, there will be more mixed farming. A new agrarianism will quite likely be based more on direct local marketing of produce than on long-distance transport. We'll also see, I think, the emergence of urban farms as lines between city and countryside continue to blur. The new agrarianism will be powered by sunlight, not fossil fuels, and agriculture will become part of a larger strategy aimed at conserving soil and storing carbon. Then, too, the new agrarianism in some way will have to acknowledge that we remain hunter-gatherers by temperament. Life on the land needs to become more fun and more adventurous than traditional farming has been.

Are there models of a new agrarianism? Gene Logsdon, for one, proposes small owner-operated farms of twenty acres or so, supported by off-farm work. Brian Donahue describes community-supported agriculture operating on a city-wide scale. The Center for Ecoliteracy in Berkeley, California, is developing gardens in public schools throughout the San Francisco Bay Area. John Todd has proposed square city blocks under glass that function as farms and wastewater treatment systems. Then, of course, there is the Amish model, still alive and healthy in small parts of Pennsylvania, Ohio, Indiana, Illinois, and Kansas. These and other agrarian practices are useful ones, however marginal their existence today in our society and industrial consciousness.

The largest barrier to a new agrarianism, then, is not the lack of good models, though of course we could use more of them, in both number and kind. It is instead the vast gap that separates sound agrarian culture from the daily lives most of us now live. Agrarianism simply doesn't compute with the experiences of people whose lives are shaped by malls, highways, television, and cyberspace. The coming generation in particular suffers a profound experiential deficit in that it is increasingly autistic toward the natural world. That deficit includes not just a physical detachment from land and food production but also a detachment from the experience of death. Until a few generations ago, birth, growth, and death were routine parts of farm experience for a large fraction of the young. On a diverse farm, some animal was being born while another animal was dying. One consequence of this lost experience: The teenage fascination with death is not transmuted by culture and lived experience into a healthy understanding of nature's ways and limits.

This deficit of experiences in nature has its social counterpart. With the dissipation and breakdown of families, many of our young have fewer good role models to supply guidance—parents, grandparents, aunts, and uncles. Many young people feel isolated, deprived of a flourishing social context in which to mature. The problem is greatly worsened by the velocity of current culture, which affects the old as well as the young. Good things often take time. There is no fast way to be a good parent or friend or, for that matter, to grow up. There is no fast way to develop good character, to sink roots, or to acquire wisdom. There is no fast way to build decent communities or to repair an eroded field.

Excessive speed helps explain another disturbing deficit apparently gaining momentum—the decline in facility with language. This decline, if real, is by no means uniform across categories of knowledge. Popular vocabulary has doubtless risen dramatically in areas having to do with sex, technology, consumption, and the corporate world. It has probably fallen sharply in areas having to do with religion, land, and practical living. The losses here are losses not only of words but also of the knowledge and experiences incorporated in and remembered by and with those words. It is a loss, too often, of all but the most rudimentary understanding of how life naturally functions and of the traditional values and restraints that undergird lasting communal life.

I know of no easy or quick way to repair what's been undone. Yet I do believe that education will have to play a critical role in fixing the damage and equipping people with the intellectual and moral wherewithal to build a better world than the one down our current path. As Wendell Berry puts it in "The Whole Horse," "between these two programs—the industrial and the agrarian, the global and the local—the most critical difference is that of knowledge." In the agrarian world, knowledge was part of the lived experience, a kind of cultural DNA that blended intuition and observation, sentiment and reason. It was knowledge tailored to a place on earth and to the peculiarities of that place. In the best of settings, it was learned as part of growing up and having to earn one's keep.

That world has largely disappeared in the case of farmsteads and farm communities that embrace true agrarian values. Too few agrarian farmers are left to pass on such knowledge, and so we are forced to ask: Who will equip us with the skills and knowledge for a twenty-first-century agrarianism?

One answer is that formal education might make up for what culture can no longer do. But our major educational institutions long ago joined the effort to industrialize, technologize, and commercialize the world, the modern farm included. They provided the analytical tools and knowledge base for an extractive economy, and they are now complicit in the effort to build a global economy in which market forces enjoy nearly free rein. Can such institutions be transformed so as to create the intellectual and practical foundations for a better world, one that blends the best of agrarianism with the best of the industrial-market system? I think so, but I'm hardly prepared to wager on it.

The transformation of education now so needed is in no way dependent on more money or more computers. What it requires foremost is a thorough rethinking of basic assumptions about what is worth knowing and how it is best learned. One piece of this rethinking has to do with the reality that pedagogy includes more than just classroom lessons and formal teaching. It also includes the vital curriculum that is embedded in campus buildings and land-

scapes. These, the places where education occurs, instruct students
in powerful ways, all the more so because they operate silently and
are rarely questioned. Consider the typical campus landscape, so
nicely managed and decorated. The fact that it resembles a country
club is hardly incidental to the financial visions of campus leaders.
What lessons does a manicured landscape convey to students?
Chiefly, it is that nature should intrude in the human realm only
when and to the degree that humans permit. Humans are in
charge; nature's role is to obey. Nature exists for humans to mold
and shape. Students who reflect on how this dominance unfolds
also pick up another message: that it's wise and proper to use fossil
fuels and chemicals to maintain human dominance, without
thought of the morrow, or fairness, or even of mammalian bio-
chemistry.

The transformation of education, though, needs to be more
Pedagogy is also embedded in the built campus environment.
Energy-inefficient buildings teach indifference to the real costs of
energy. Toxic and hazardous building materials tell students that
it's fine to be oblivious to the dark ways in which we write our presence in the world. At Oberlin College, my home for the past
decade, we have several buildings designed by a company that built
prisons for the state. With a bit of razor wire and bars for the windows, they would still make good prisons. This, too, is a pedagogi-
cal practice. And it is in such places that we purport to encourage
creativity, intellectual growth, and responsibility.

The transformation of education, though, needs to be more
thorough still. For a long time, we've regarded rigor as a matter of
going deeper and deeper into an ever more specialized subject, aiming, as someone once put it, to know more and more about less and
less until one knows everything about nothing. This type of rigor
has its place, but it is a dangerous rigor in that those who embrace
it can easily lose sight of why they work, how their products will be
used, and even whether they're doing harm or good. The industrial
world is fueled by rigor of this type, and many of its ills are caused
by it.

There is, however, another view and form of rigor, one that entails the facility to think, as Aldo Leopold put it, "at right angles" to one's field of specialization. Let me illustrate. Every spring, I teach a course in sustainable agriculture. In it, I often encounter bright biology majors who know how genetic engineering is done but who can hardly begin to think clearly about the wisdom and morality of reweaving the fabric of life. They've learned to be unskeptical, which is to say unscientific, about the craft of science itself. This is the worst kind of naive positivism, and it places a great deal in jeopardy. We need a science of a more integrative type, one that connects knowledge from different fields and perspectives. We need, as well, more knowledge of the type that comes from a sound connection between intellect and emotion. In the words of Vine Deloria:

> In most respects, we have been trained to merge our emotions and beliefs so that they mesh with machines and institutions of the civilized world. Thus many things that were a matter of belief for the old people have become objects of scorn and ridicule for modern people. We have great difficulty in understanding simple things because we have been trained to deal with extremely complicated things.

One trademark of the agrarian mind was precisely this ability to integrate. The agrarian farmer couldn't till the land by drawing on a single scientific discipline, nor by using only verified knowledge. Intellectual breadth was a matter of necessity. Intuition and tradition were instinctively mixed with empirical knowledge. Responsibility was undivided.

The marriage of the agrarian and industrial minds will require the merger of these different ways of knowing—the empirical, specialized world of science and the holistic, moral, experiential world of the responsible agrarian.

Young people, in my experience, have an intuitive sense of this need for integration, even if they can't explain it. Every spring, I take that same sustainable agriculture class to an Amish farm in Holmes County, Ohio. The students' reaction is overwhelmingly one of surprise, particularly at the joy and satisfaction the Amish display and at the creativity and intelligence required to farm well. Most students have never been on a farm, and they assume that farmers are dull, humorless, and overworked. As they leave this farm, their typical comment is "This is the best day I've had," or "Those people seem so real!" Reactions such as these reveal a hunger for something deeper than what students typically get in college. They also reveal, I believe, that many students yearn for lives rooted in deeper soil, however much we might hear about their zeal for money and careers.

How can those of us who bear the label "professor" offer this kind of education? How might we instill in our students an awareness of the land community? How can we instill affection for community and the practical competence necessary to restore land to health? I cannot offer full answers to such questions, but I can relate some of the experiences I've had.

In 1995–1996, Oberlin College began the design and development of the Adam Joseph Lewis Center for Environmental Studies. From the beginning, the design phase was understood as an educational process, involving students, faculty, staff, and the general public. Thirteen public design charrettes were undertaken. From them many goals emerged, including the hope that the building could purify its wastewater on site. To meet that goal, we called in engineer John Todd to design a water purification system. He calls his designs "living machines," and they resemble tropical greenhouses. A second design goal was to power the building insofar as possible by current sunlight rather than ancient sunlight stored as fossil fuels. That, too, was an ambitious goal in a region that has as many cloudy days as Seattle has. The current-sunlight target forced us to maximize energy efficiency—cutting usage to one-fourth of the energy consumed in typical buildings—even while we included numerous windows so we could enjoy daylight. We now realize that such a target can be achieved early in the life of a building; indeed, even in a cloudy, cold region, a building such as ours could operate at or near the point of being a net energy exporter. The energy system for the building is a 3,700-square-foot, roof-mounted photovoltaic array capable of producing 75,750 kilowatt-hours per year, or 18,700 Btu's per square foot—about the energy used in six or seven normal homes.

A third design goal proposed that the Lewis Center be able to "learn" over time in the sense that it could incorporate future technology and not be restricted to technology in existence during its design. The result is a high degree of flexibility and modularity throughout the structure, enough to accommodate future technological changes. A related standard called for the application of what architect William McDonough calls "products of service." Any building is, in effect, a conversation between two different metabolisms—one industrial, the other ecological. The difference between the metabolisms is easily expressed: If a building component would make good soil when discarded, it is part of the ecological metabolism; if it would not, then it is part of the industrial metabolism. Industrial components go back to the manufacturer to be remade into new products, both to save raw materials and to avoid contaminating land. In our building, the best example is the carpet leased from Interface, Inc., which will be returned eventually for refashioning into new carpet. It has been designed for reuse in the manufacturing process, thereby closing the loop in the industrial metabolism in a manner that reflects the agrarian mind's respect for intact cycles. The lesson here is simple: The two systems should not be commingled.

The governing standard for the Lewis Center was what we deemed a deeper, more moral kind of aesthetic. We aimed to cause no ugliness, human or ecological, whether somewhere else or at some later time. That standard forced us to think "upstream," to the wells, mines, forests, and manufacturing plants where resource flows begin, as well as "downstream," to the effects our building will have on future climate, biodiversity, and human health. As in the case of all high standards, our achievement has been a matter of degree. Nonetheless, the standard forced us to reckon with things few designers and builders take into account. All wood in the building, including laminated beams, roof decking, plywood, and trim lumber, came from forests managed for long-term ecological health. Nontoxic paints and finishes were used throughout.

No architect could design a project such as this alone. For that reason, we asked William McDonough + Partners, PLC to work with a design team that included John Todd, Amory Lovins, Bill Browning, Ron Perkins, Adrian Tuluca, Andropogon Associates, Ltd., John Lyle, Steven Strong, scientists from the National Aeronautics and Space Administration, and a dozen others. Many of these will participate in the second phase of the project, which will include refining the building's energy performance and designing a demonstration regenerative fuel cell to convert solar-generated electricity to hydrogen and then back to electricity when needed.

Buildings such as the Lewis Center are not just ends in themselves but also means to larger ends. In our case, the success of the project must be measured in part by how much it has improved the thinking of Oberlin students. It was designed not only as a place for classes but also as a tool to engage students in ongoing operations in ways that will enable them to learn. Wastewater management (ecological engineering), for example, is now part of our curriculum. The design and operation of renewable energy technologies is also coming into the curriculum as a blending of high technology and good design.

The marriage of agrarian and industrial minds is perhaps even more evident in the landscape around the Lewis Center, with its orchards, gardens, meadow, forest, pond, and native plant wetland. This is not a decorative landscape but one intended to combine land management with skills of gardening, orcharding, horticulture, and species preservation. In the second phase of the project, we plan to expand the orchard and build a greenhouse and laboratory. Our aim overall has been to arrange a setting in which students, themselves mostly products of the urban world, can be reconnected to the Creation in a practical way. Two miles to the east, we are developing a seventy-acre farm site as a home for a community-supported farm begun in the mid-1990s. That project will be a working, for-profit enterprise to grow food for the college and community while providing a site for public education.

In much of our work at Oberlin, we've found ourselves promoting what can be regarded as agrarian ends while using different kinds of analytical skills, such as full-cost accounting and least-cost, end-use analysis, as well as drawing on high-technology tools such as geographic information systems. For example, we have worked with the Rocky Mountain Institute to determine what would be required for the entire college to become "climatically neutral"—effectively powered by sunlight—by the year 2020. That assessment required a systems analysis of institutional energy use and rigorous application of least-cost analysis and technology forecasting.

One obvious question about our endeavors is whether they have much relevance to urban areas. Oberlin, after all, is a small town in rural Ohio, and land around it does not carry urban price tags. Can the lessons we learned help deal with urban problems in ways that might make our cities sustainable? Might they help blur the line between urban and rural by bringing agrarian ideals and practices to the urban scene?

We believed they could, and we sought to test that belief by trying to transfer our work to nearby Cleveland. We initiated a speaker series, funded by Cleveland corporations and The George Gund Foundation and led by a recent graduate of our Environmental Studies Program. That effort helped ignite a region-wide discussion about the application of ecological design to our part of the old rust belt. One tangible result: the formation of a "green building coalition" of architects, engineers, developers, county planners, public officials, and nonprofit groups that aims to change the face of the region. Projects under way now include an effort by EcoCity Cleveland to build an environmentally designed community along the rapid transit line; ecological redesign of a bank building in a transitional neighborhood; the makeover of a brewery aiming to become a zero-discharge company; and ecological restoration of the Cuyahoga River valley.

It is time for us, plainly, to be asking larger questions, developing bolder visions, and thinking in terms of centuries, not years. Can we put agriculture back into an urban fabric? Can we develop businesses and entire communities that produce no waste? Can we power buildings and communities by sunlight? Can we incorporate the frugality, ecological competence, celebratory spirit, and neighborliness of rural life at its best with the dynamism, wealth, and excitement of the city?

I believe that we can.

Do the examples I've described constitute an adequate answer to the problem? Hardly. But along with other examples, here and elsewhere, they do provide beginning points for the transition to a better world. If such a world is to become manifest in the decades ahead, it will require the marriage of urban-industrial and rural-agrarian perspectives. As in traditional agrarian communities, sunshine must be the primary power source. As with well-operated traditional farms, waste must be eliminated through the combination of better design and frugality. Like rural people generally, citizens must become ecologically competent enough to meet many of their needs with local resources. That world must display the kind of neighborliness and affection for place characteristic of good rural communities. Yet, on the other side, it needs to embrace the dynamism, tolerance, and larger worldviews that we mostly associate with urban life at its best.

Can we create that kind of world? The limiting factor, I believe, is not so much resources or money as imagination. In the same way The Land Institute has helped create a vision of ecological agriculture, colleges and universities might pioneer the work necessary to create a compelling vision of an ecologically sustainable, spiritually sustaining world. We have it on high authority that without such a vision, we will perish.

Not long ago, a foundation with which I'm associated helped support the conversion of a school parking lot into a large garden. The garden was intended to produce food for the school cafeteria and to yield a surplus to sell to local restaurants. As planned, much of the work was done by young students, and afterward we asked several of them to present the project to the foundation board. When asked what they'd learned from the effort, one enterprising, eloquent boy offered a surprising response: "Asphalt isn't forever."

Now, if a young student can grasp this truth, perhaps others of us can see that decaying, sprawling, crime-ridden cities are not forever, any more than degraded rivers, eroded fields, abandoned farms, and dying small towns are forever. Better alternatives are available. It is up to us to bring them into being.

PART II

THE SIRENS
OF CONQUEST

Chapter 8

The Decline of the Apple

Anne Mendelson

Perhaps no food crop is more entwined with American folklore than the apple, so it is fitting that in the story of the apple—told here by food historian Anne Mendelson—we see at work the powerful forces that have redirected America's chief means of obtaining food and relating to land. From a crop widely grown by households everywhere, the apple gradually became a specialty item, its production dominated by large growers. Breeding goals shifted from taste and nutrition to superficial characteristics such as appearance, shape, and ease of shipping. Growers specializing in apples became vulnerable to the price fluctuations of a single market. And as production shifted from dispersed farms and backyards to monocultural groves, disruptive chemicals were needed to ward off pests.

Although Mendelson's narrative dwells on discouraging long-term trends, she perceives more favorable signs as she brings her story up to the present. Home orchards are making a modest comeback, and consumer demand is stimulating the revival of many heirloom apple varieties.

This article is loosely based on an earlier version that formed part of a talk given by Lee Grimsbo and me to the Culinary Historians of New York in June 1988. I am grateful to Mr. Grimsbo for sharing with me not only important books but also his wide knowledge of American pomiculture. I have omitted, with regret, several critical developments in American fruit-growing history. The most important are the worldwide export trade in American fruit from about the 1850s into the early twentieth century, the early golden age of scientific pomology beginning with the establishment of the state agricultural experiment stations in 1887, and the rise of West Coast fruit farming (with corresponding loss of eastern production) from about the turn of the twentieth century.

The studies of nature have been wisely ordained by Omnipotence as the most pleasing to the mind of man; and it is in the unbounded field which natural objects present, that he finds that enjoyment which their never-ending novelty is peculiarly calculated to impart, and which renders their study devoid of that satiety which attaches itself to other pursuits. Most wisely has it thus been prescribed, that by an occupation of the mind, itself inviting and recreative, we should be insensibly led on to a development of the intricacies of nature, and be thus taught to appreciate the beneficence of the Creator, by a knowledge of the perfection and beauty which mark the labours of his hand.[1]

This exalted tribute sounds for all the world like some minor Enlightenment philosopher explaining the essential harmony of religion and natural science. In fact, it is from a manual on successful fruit growing written by a nurseryman in Flushing, New York, during the administration of Andrew Jackson. It asserts something that struck chords in many American hearts for several generations: that cultivating an orchard offers intellectual, even spiritual, benefits befitting an enlightened citizenry. True, the writer also noted that the climate and soil of the United States had proved wonderfully suited "to the culture of the various fruits" and that such pursuits "tend greatly to advance the wealth of a community." But what roused him to real eloquence was the delight an orchard yielded to the contemplative mind.

Had William Robert Prince, the author of *The Pomological Manual,* been transported from 1831 to 1914, he would have found the Apple Advertisers of America putting their muscle behind National Apple Day with a campaign including prizes for school essays on apples and free distributions to florists or restaurateurs willing to help plug the product. Getting the public to buy apples, the organization explained in its new publication *The Apple World,* was nothing to be left to chance. "We are finally waking up to the fact that the Apple business is too big, too important, too valuable, to be allowed to drift about on the commercial sea 'the sport and prey of racking whirlwinds.' We must steady its course through selling efficiency made possible by sane and sound publicity methods."[2] The cover depicted a large globelike apple afloat on cosmic clouds, with a map of North America firmly plastered across the center beneath a label reading "The World's Apple Orchard."

In the gap between these two visions of the American orchard lies a story of hope and disillusionment unmentioned in most histories of American agriculture. The twentieth-century end of the story is familiar: a society that wouldn't eat at all if not for steep, risky capital investments on the part of growers continually struggling for a surer hold on markets. But at the other end, close to the birth of the republic, is something we have mostly forgotten: The vast resources of the continent then seemed to promise that ordinary people might not just lead self-sufficient lives on their own land but enjoy a reward that once had been only the pleasure of the rich—an orchard.

Most of us vaguely recall Jefferson's hope to see America a nation of small farmers. It is less widely understood that, even in Jefferson's lifetime, others were refining the plainspun agrarian ideal, to suggest that in a land of equals, the smallest farm should also make room for the finer things of life. Gardens, especially orchards, were longstanding symbols of sheltered leisure and privilege. More recently, they had become the venue for such glamorous attempts to improve on nature as the seventeenth-century tulip mania or the practice of "inoculating" (grafting) fruit trees, deplored in Marvell's poem "The Mower Against Gardens." By the end of the eighteenth century, the luxury of a flower, vegetable, or fruit garden had filtered down more widely to the middle class. So had a conviction that such things, far from being fripperies, were good for the mind and spirit.

We can put together a lively picture of the American orchard idyll and its nineteenth-century fortunes by looking at accounts of orchard fruits—particularly apples—in the wealth of horticultural writing already well established by midcentury. Today we rarely think of classifying fruit and vegetable raising under the rubric of "horticulture" (garden cultivation) as opposed to "agriculture" (field crop raising). But, for some years of our history, the word "horticulture" had a novel sheen, promising an America made happier and more beautiful by the civilizing influence of gardening skill in every rural home.

Nothing seemed to promise more happiness and beauty than the apple. It has always been the American fruit par excellence, partly reflecting an early preference for cider as a beverage. Probably no country on earth offers more extraordinary scope in terms of growing conditions to the genetic possibilities of *Malus pumila,* which throws up a new combination of qualities every time an apple blossom is pollinated. Amelia Simmons's *American Cookery* declared in 1796 that, if every American family had a properly cared-for apple tree, "The net saving would in time extinguish the public debt, and enrich our cookery."[3] This patriotic thought only echoed a widely shared faith in the virtues of fruit raising. Already some of the agricultural societies founded in the eastern states during the eighteenth century were devoting much attention to horticulture; they would soon be supplemented by horticultural and later pomological societies singing the praises of orchards. An early paean from this era was delivered around 1804 by a member of the Kennebec (Maine) Agricultural Society:

> When we consider the various manners in which fruits are beneficial, when we recollect the pleasure they afford to the senses, and the chaste and innocent occupation which they give in their cultivation; when we consider the reputation which they communicate to a country in the eye of strangers, especially as

affording a test of its climate and industry; when we remember the importance of improving the beverage which they are intended to supply; when it is calculated under how many solid forms they may be exported (as dried, baked, and preserved, as well as in their natural state;) and lastly, when we reflect upon the utility of giving to our rural labours a thoughtful turn, which is the best substitute we have, after having quitted our primeval state; I say, when we consider these things, it will appear that the subject of fruits, which were the first earthly gift of Providence to man in his more favoured state, may well continue to merit both the publick and individual attention.[4]

James Thacher's 1825 treatise *The American Orchardist,* which approvingly quotes this lofty thought, notes that, "in the whole department of rural economy, there is not a more noble, interesting, and beautiful exhibition, than a fruit orchard, systematically arranged, while clothed with nature's foliage, and decorated with variegated blossoms perfuming the air, or when bending under a load of ripe fruit of many varieties."[5]

One could multiply examples from the early American horticultural literature. Like William Robert Prince, the commentators duly noted the commercial potential of the product. But what is striking today is their moral and aesthetic fervor and the assumption that fruit raising was a pleasure by which any citizen could enrich our common life. That note continued to be sounded as the canals and railroads began opening up more of the continent to cultivation. It was still being sounded late in the nineteenth century. One strain of opinion thought home-scale pomiculture beneficial to the family, viewed by most nineteenth-century writers on domestic affairs as the great moral nurturing ground. In the *American Fruit Culturist* (1849), John Jacob Thomas declared that the home with an attraction like its own fruit garden had a "salutary bearing" on growing children. John A. Warder's *American Pomology: Apples,* published in 1867, also asserts that a lifetime of happy, improving associations can be stored up through childhood by the delights of the home garden and orchard. Beyond the strengthening of familial ties, Warder thought that planting and taking care of one's own tree was an excellent early lesson in property rights. Fruit stealing was a common symbol of juvenile crime, nineteenth-century style, but this observer saw the cure not in deploring the crime-producing or "sybaritic" effect of "luxuries" like fruit but in letting young children learn about *mine* and *thine* from growing their own. Each would then "appreciate the beauties of the moral code, which he will be all the more likely to respect in every other particular."[6]

An allied claim was that small-scale fruit growing fostered a larger sort of communal unity and stability. The well-known Rochester nurseryman Patrick Barry argued along this line in *The Fruit Garden* (1852). Fruit culture, he announced as a self-evident truth, was an interest linking all classes and occupations from "agriculturist" to "merchant or professional man" and "artizan."

> It is the desire of every man, whatever may be his pursuit or condition in life, whether he live in town or country, to enjoy fine fruits, to provide them for his family, and, if possible, to cultivate the trees in his own garden with his own hands.... Fortunately, in the United States, land is so easily obtained as to be within the reach of every industrious man; and the climate and soil being so favorable to the production of fruit, Americans, if they be not already, must become truly "a nation of fruit growers."[7]

Several years before, a farmer named Henry French in Exeter, New Hampshire, had praised the stabilizing effect of apple growing in a letter submitted to the Patent Office for its 1849 agricultural report. (From 1839 until President Lincoln established the

Department of Agriculture in 1862, the Patent Office acted as a general clearinghouse of information on all aspects of farming.) French opined:

> An influence is much needed in New England to counter-balance the roving propensity of her people; an influence which is nowhere so surely to be found as in the strengthening of home-ties by the union of labor with the works of nature. He who has planted a tree, will he not desire to eat of the fruit thereof? and he whose father has raised it, will he not feel it to be almost sacrilege to give it into the hands of strangers?
>
> Patriotism has no basis so secure as in the love which man has for his home and the home of his fathers.[8]

The force of this claim cannot be appreciated by casual visitors to a modern orchard. Most apple trees are by nature slow to mature. What Henry French knew as standard apple trees reached spreading heights like twenty or even forty feet and might take from seven to twelve years to start bearing, though they then usually went on for decades. A long-lived, space-hungry orchard like that obviously gave its planters a stake in staying put—something altered in our lifetime not only by urban sprawl but by the practice of grafting virtually all varieties onto dwarfing or semidwarfing rootstocks that reduce the size and life span of the mature tree along with the time it takes to start bearing. (A modern orchard may crop within two or three years.)

If we want an eloquent summary of the inspiration that the nineteenth-century pomologists sought in fruit raising, it is furnished by John Warder in a passage insisting that, for all the importance of agriculture, it is the pursuit of horticulture that "always marks the advancement of a community":

> As our western pioneers progress in their improve-

ments from the primitive log cabins to the more elegant and substantial dwelling houses, we ever find the garden and the orchard, the vine-arbor and the berry-patch taking their place beside the other evidences of progress. These constitute to them the poetry of common life, of the farmer's life.[9]

But the Exeter farmer, noting that New Englanders tended to *leave* New England, had hinted at something already quite unlike the manual writers' picture of the happy home orchard as a bright influence cementing family and community. The fact was that the life regularly painted as American rural destiny had little to do with the actual choices of people living on the land, even in the first heyday of American orchardry. Home orchards indeed there were, for a very long time. One still sees their remnants on old farms in many parts of the Northeast and the Midwest. To travelers, they seemed one of the most characteristic features of the American landscape. Sereno Todd's *The Apple Culturist* (1871) quoted Horace Greeley on the "air of comfort and modest thrift" that such living tokens of refinement gave to a homestead:

> If I were asked to say what single aspect of our economic condition most strikingly and favorably distinguished the people of our Northern States from those of most if not all other countries which I have traversed, I would point at once to the fruit trees which so generally diversify every little as well as larger farm throughout these States, and are quite commonly found even on the petty holdings of the poorer mechanics and workmen in every village, and in the suburbs and outskirts of every city.[10]

These pretty clusters of trees barely hinted at what had happened since the early horticultural cheerleading in the first decades of the century. The fact was that thousands of farmers had been persuaded to invest great amounts of land and money in fruit, especially apples, *as a specialized cash crop.* "The public mind has waked up to the importance of this subject, and in some sections is roused even to a sort of enthusiasm upon it," a Massachusetts contributor to the 1849 Patent Office report sanguinely noted.[11] Three years later, in *The Fruit Garden,* Patrick Barry grandly pooh-poohed those who thought the fruit-growing craze "a sort of speculative mania."[12] Ominous realties were masked by this optimism. The true story of nineteenth-century pomiculture is how the business of producing apples for a market assumed a demanding, risky life of its own even as charming tributes to home orchardry filled the pages of manuals and journals. We see the two unfolding together in the preface of the foremost pomological manual of the century, A. J. Downing's *Fruits and Fruit Trees of America,* first published in 1845 and issued in many revisions until 1896:

> America is a *young orchard,* but when the planting of fruit trees in one of the newest States numbers nearly a quarter of a million in a single year; when there are more peaches exposed in the markets of New York, annually, than are raised in all France; when American apples, in large quantities, command double prices in European markets; there is little need for entering into any praises of this soil and climate generally, regarding the culture of fruit. In one part or another of the Union every man may, literally, sit under his own vine and fig tree.[13]

The vine-and-fig-tree model was just what the early horticulturists had preached as a fit image of rural America. But it was not consistent with the vision of orchards as gold mines. Downing's claims about European prices for American apples and the New York markets' selling more peaches than the entire output of France, at a time when hopeful investors were putting in a quarter of a million fruit trees in *one* new state alone, suggest a widespread rush to make a buck. It was a dangerous game even then, for circumstances that would be minor disappointments or even blessings to the self-sufficient rural family with a plot of fruit trees were absolute catastrophes to a farmer who had staked all on a large orchard.

One occasion of rude awakening was the seasonal glut of the market, still worse in years of bumper crops. This was, ironically, the consequence of living in a peerless region for raising apples. At the time of the initial fruit-growing boom—the last twenty or thirty years before the Civil War—growers competing with dozens or hundreds of others for city fruit markets could do little to store apples more than a few weeks, except for varieties like Winesap that naturally kept well over the winter. Some of the best dessert apples, articles that commanded glamorous prices in the cities, were fragile and short-lived, not easy to get to market in good condition under the best circumstances. (The Fameuse or Snow apple, thought to be a parent of the McIntosh, was one of these difficult beauties.) Another problem was the biennial tendency of some popular varieties including Baldwin, Roxbury Russet, and Northern Spy, which might bear handsomely one year and hardly at all the next.

Turning large tracts of land in New England, New York, and the upper Midwest into fruit orchards had another effect: It created ideal habitats for insect scourges that had been less devastating in a time of scattered, nonintensive fruit cultivation. These pests now assumed dire economic significance. From the early boom in the 1830s until near the end of the century, orchardists regularly lost whole crops (or large percentages) to insects. The nineteenth-century growth of the city "fancy trade," which paid premium prices but required cosmetically attractive fruit, made even slightly blemished apples less acceptable than they had been before the rise of commercial orchardry. For many years farmers were essentially helpless against insect damage, though they continually wrote to the horticultural journals with tips like the mixture of whale oil

soapsuds and tobacco water tried by many against a weevil called
the plum curculio, which also relished apples. One reader of A. J.
Downing's celebrated magazine *The Horticulturist* reported in 1847
that he had "found that the worm would live after having been
immersed in tobacco water so strong as to be as dark as port
wine."[14]

The solution to this problem did not appear until the 1880s,
when the first effective pesticides applied by spraying were introduced and at once became one of the unavoidable fixed costs of the
orchard business. Effective storage was another expensive boon.
Neither ice cooling (for which there was fairly good technology
from the 1850s and 1860s) nor mechanical refrigeration (well
advanced by the last decades of the century) was feasible for most
growers to use on their own farms. It was primarily merchants and
distributors who would profit from these advances in storage,
which incidentally jacked up the cost of getting the fruit from
farmer to consumer.

Almost as soon as the big commercial fruit-raising push had
reached the northern states, it was obvious that many people had
not landed themselves in the vine-and-fig-tree life or in a fountain
of riches. From the viewpoint of modern connoisseurs, the scram-
ble for a competitive edge had at least one attractive upshot. This
was the great proliferation of apple varieties, something that would
not have been possible during the centuries when most apple trees
were chance seedlings and only a few hobbyists were interested in
"inoculation." (No apple variety can be truly propagated by anything except grafting or budding—"inoculating," as these practices
usually were known until the nineteenth century.) A few dozen
new American varieties had been propagated, mostly in New York
and New England, during the eighteenth century. At least two of
these—the Esopus Spitzenburg and the Newtown Pippin—continued to be regarded as the absolute top of the line until well past
1900. But the nineteenth century saw a virtual deluge of new varieties. Suddenly the immense genetic variability of the species was

understood as an opportunity for profitable experiment. Already in 1852 Patrick Barry estimated the number of cultivated apple varieties to be at least a thousand, though he chose only 150 for mention in *The Fruit Garden*.

From the 1830s and 1840s on, orchardists developed periodic enthusiasms for some new wonder apple. The Northern Spy and the Jonathan were early sensations that happen to be among the best-flavored apples anywhere on the planet. But good flavor was not the prime consideration of everyone who watched a new seedling for signs of promise and added it to the flood of propagated varieties. Though Barry counted "every [variety] of *real excellence* as an additional blessing to the fruit growers and to society, for which they should be duly grateful,"[15] these blessings were not equally shared among all growers. An apple touted as a miracle in central Connecticut might be mediocre or almost ungrowable in Ohio, Michigan, or western New York. A farmer might find this out after an investment of ten years' time and dozens of acres newly carved from wilderness. Of the literally thousands of varieties grown on some scale during the nineteenth century, only a few dozen proved to have wide commercial potential, and their survival did not necessarily depend on quality. By the end of the Civil War, many growers saw great merit in so-so apples with some particular selling point like size, bright color, smooth skin, inoffensive sweetness, highly reliable cropping, toughness and transportability, or timing to hit the dead spots of the seasonal market.

Some canny observers shortly began to think that the future belonged to a pomiculture based on *fewer* kinds of apples, not more. (It hardly needs saying that they were right. Today most Americans have never seen more than seven or ten varieties of apple, while some like Baldwin and Greening, considered mediocre a century ago, are now head and shoulders above most of what we get.) An orchardist named Seth Fenner, who delivered a paper to an audience of western New York fruit growers in Oswego in 1886, saw quite correctly that the great wealth of apple varieties then being cultivated was a mistaken route to the sacred purpose of making money. "The one great mistake of American orchards," he philosophized, "is the multiplicity of varieties. Multiplicity of varieties works as great damage to orchards as polygamy does to the Mormons, and you want to avoid it. Select a few standard varieties, especially for market."[16] He suggested a list of nine practical choices that entirely omitted the four great glories of New York apple growing—Newtown Pippin, Esopus Spitzenburg, Northern Spy, and Jonathan.

Well into the orchard gold rush, some observers were putting the best face on what had developed into a grinding struggle for many. Allen W. Dodge, the Massachusettsan who had hailed the fruit-growing boom in the 1849 Patent Office report, clearly had heard apprehensive assessments in some quarters. "Suffice it to say, it is no speculative or visionary scheme," he insisted, "but a safe and permanent investment that will yield golden dividends, so long as our soil and seasons shall continue to be as propitious as they have heretofore been." Scorning suggestions by those of "narrow vision" that the market would soon be glutted, he added in a footnote that, anyhow, the prospects of the European market for American fruit were limitless.[17] Patrick Barry's *Fruit Garden* enumerated still more promises of great things to come:

> At one time apples were grown chiefly for cider; now they are considered indispensable articles of food. The finer fruits, that were formerly considered as luxuries only for the tables of the wealthy, are beginning to take their place among the ordinary supplies of every man's table; and this taste must grow from year to year, with an increased supply. Those who consume a bushel of fruit this year, will require double or treble that quantity next. The rapid increase of population alone, creates a demand to an extent that few people are aware of. The city of Rochester has

added 20,000 to her numbers in ten years. Let such an increase as this in all our cities, towns, and villages be estimated, and see what an aggregate annual amount of new customers it presents.

New markets are continually presenting themselves and demanding large supplies. New and more perfect modes of packing and shipping fruits, and of drying, preserving, and preparing them for various purposes to which they have not hitherto been appropriated, are beginning to enlist attention and inquiry.[18]

Dodge and Barry were right in predicting a vast expansion of markets through new consumption patterns, European exports, experimental technology, and sheer population growth. They were wrong in thinking that gluts could not occur or that an expanded orchard capacity could easily surmount any market vagaries. In some ways the future was bright for American orchardry. It was not bright for the self-sufficient family-scale idyll that continued to be widely depicted as American rural life. The reality of the situation appears much more clearly a dozen-odd years later, in an article on horticulture contributed by an observer named M. L. Dunlop to the transactions of the Illinois State Agricultural Society for 1865–66:

When we take a view of the great variety of agricultural products, we can come to no other conclusion than that it must needs have a division of labor similar to that of the mechanic arts. The farmer may raise grain and cattle and fruits and garden vegetables for market, but when he attempts them all on an even scale, he will make a signal failure. He must choose one of these as the main business, and make the others only secondary, to be pursued as convenient. This is becoming a well settled principle in rural economy, and the true one. We have grain farmers, dairymen, stock-growers, hog-raisers, orchardists, and market-gardeners. . . .

We have reached that point in our progress that the demand for horticultural products is of such a nature that a large part of our rural population and capital can be most profitably employed in it. The demands from our manufacturing cities and marts of commerce must be supplied. But we now have new demands upon us—the railways have opened up to us new markets, and we can send the products of the orchard to the miners of the Rocky Mountains, and the regions of the more inhospitable North.[19]

The change was upon the farmers of the eastern and midwestern fruit-growing regions before anyone had grasped its implication: What had been the pleasurable and spiritually rewarding pursuit of horticulture as a blessing of democracy would have to become just another branch of a very specialized, professionalized, segmented American agricultural economy in which no effort was free of real financial risk. The farmer must decide not what he wanted to grow but what he wanted to sell.

This candid recommendation of specialized fruit farming is a long way from William Robert Prince's praising orchardry as a refreshment for mind and soul. The second great spurt in American pomiculture—ushered in by real scientific advances in pomology around 1890 and lasting until after World War I—would travel even further from the visions of plenty, equality, and enlightenment once conjured up by the thought of raising one's own fruit. Even in the 1860s and 1870s, unhappy accounts of growers' tribulations suggested that what many were really casting about for was the promised cure-all hailed in 1914 by *The Apple World*—advertising, a kind of camshaft linking the crop in the orchards with desired levels of consumption.

Today, tasting the wretched apples from a supermarket or touring an irrigated factory-scale orchard, one can scarcely imagine that people ever attached moral or spiritual value to such things. And in hindsight it is clear that the Eden seen in the American orchard by the horticultural manual writers never really existed. But there is another postscript to the story of fruit growing as part of the early American democratic vision.

In a sense, something that can be called the humanist side of fruit raising never really died out. It was kept alive by some pomologists well into the age of marketing pitches. In 1922 the great horticultural writer Liberty Hyde Bailey commented that a current industry campaign "to lead the people to eat an apple a day" was beside the mark: "To eat an apple a day is a question of affections and emotions." He meant not trivially manipulated preferences but the real inner loyalty to ideals of excellence that distinguished "a good amateur interest in fruits."[20] Seven years before, the same observer had sadly noted what he saw as a true cultural loss:

> It is much to be desired that the fruit-garden shall return to men's minds, with its personal appeal and its collections of many choice varieties, even the names of which are now unknown to the fruit-loving public.... The commercial market ideals have come to be controlling, and most fruit-eaters have never eaten a first-class apple or pear or peach, and do not know what such fruits are; and the names of the choice varieties have mostly dropped from the lists of nurserymen. All this is as much to be deplored as a loss of standards of excellence in literature and music, for it is an expression of a lack of resources and a failure of sensitiveness.[21]

Bailey was more right than he could have foreseen about the dismal effect of "commercial market ideals" on the quality of American fruit. But, since he wrote, many thousands of people have done much to justify his regard for "the amateur, who, as the word means, is the lover."[22] For some decades of the twentieth century, the fruit-growing tradition found its best American expression in generations of Italian and Middle Eastern immigrants who would no more have been without fruit trees in the smallest yard than without shoes or a roof. And, in the last few decades, another constituency has begun restoring something that belongs to all of us by searching out "heirloom" varieties of apple—sometimes obscure local specialties—unknown to commercial orchardry for fifty or eighty years. There are enough of these Quixotes seeking the poetry of common life to be recognized as a market by some of the largest commercial nurseries. As a result, the outfits that supply the big growers with the makings of edible sawdust now also offer home orchardists an increasing selection of apples fit for William Robert Prince's "occupation of the mind, in itself inviting and recreative." With no idea on earth except to enjoy a heritage of excellence first developed on American soil and linked with one of the better American aspirations, hobbyists are bringing back to life apples like Fameuse, Swaar, Wagener, Esopus Spitzenburg, Roxbury Russet, and Westfield Seek-No-Further. There is something in this effort that transcends mere hobby. It keeps alive a loyalty to a younger America that Professor Bailey understood sixty-seven years ago when he called the pomological amateur "the embodiment of the best in the common life, the conservator of aspirations, the fulfillment of democratic freedom."[23]

Chapter 9

The Competitive Edge

Don Kurtz

This selection is drawn from the novel South of the Big Four, *a penetrating agrarian critique of industrial farms and farm life at the end of the twentieth century. It centers on the fictional community of Delfina, Indiana, which has become dominated by large-scale, chemical-dependent methods of grain production. Agrarian values are barely discernible as industrial culture approaches its zenith.*

The narrator of the novel is Arthur Conason, a thirty-year-old who had drifted away from his family's farming operation in Indiana to work for a dozen years on the ore boats of Lake Superior. He returns to his family's abandoned farmhouse—one of the region's many marks of decline—searching aimlessly for moral grounding and a new start. Arthur is hired as a farmhand by Gerry Maars, a prominent practitioner of the get-big-or-get-out school of farming. Although superficially wealthy, Maars lives on borrowed money, acquiring land, equipment, a big house, a boat—all in an unchecked attempt to succeed and make plain his success. So frantic is his work that he cannot enjoy his leisure, and he has become coarsened to aesthetic and religious concerns. Only indirectly do we gain a sense of how his farming is degrading the land.

So long and so thoroughly has the competitive spirit sapped Delfina that dealings among residents are inherently suspicious and town youths suffer the too-common signs of alienation and despair. Maars continues to labor as a community leader and openly laments lost communitarian ties, yet he throws himself headlong into the competitive race that is their undoing. Arthur displays similar contradictions: Even though pained by the loss of old ways, he outwardly denies their importance; even as he

129

*admires Maars' energy and success, he sees as clearly as anyone the social
ruin these are helping bring on.*

  *Don Kurtz grew up in east-central Illinois in and around farm com-
munities. While working on the book, he spent several seasons as a farmhand on a large cash-grain operation.*

We'd still called it the Bremer place when Gerry [Maars] and I
talked, even though he'd lived there almost twenty years himself,
before [he and his family] moved into town. Town was where most
of the farmers had ended up, the ones who'd prospered anyway:
Charlie Sellars had settled in with his widow up at the Lake, and
Wendell Bardacke had a big house at the far end of a pasture he'd
managed to convince people was a subdivision. Gerry Maars had
his own showplace, complete with backyard swimming pool, in
Forest Meadows, on the west side of town. I'd driven by a couple of
times since I'd been back, and the next morning went by again, on
my way into the truckstop for breakfast. His black Fleetwood was
out in the driveway. It was gone on my way back, so I stepped on
the gas—if he was anything like my dad, he'd say nine just so he
could be out there at eight-thirty, complaining how nobody in the
world wanted to work anymore but him.

  At the Bremer place two tractors were pulled up in the barnyard. One was a big International, with dual tires mounted under a
huge glassed-in cab. That was standard enough by then, and so was
the planter, eight rows with a central seed drum the size of a bath-
tub. The other tractor was the flagship, seventy-five thousand dollars if it was a dime, a big 4586 four-wheel drive. It rose up above
me huge as a house, hooked to a twenty-foot disk.

  Since nobody was there yet, I walked back through the sheds.
His trucks had been repainted over the winter, so even in the shadows they shone like new. He'd been quite a livestock man at one
time, but the hoghouses and pens were gone. In their place a half-
dozen grain storage bins stretched out along the lane, hooked up by

chutes and augers to a long orange grain dryer. In the shop he had a fine set of tools.

It was a subtle kind of thing, but right there to see if you looked: the very best money could buy. Regular old gravity wasn't even progressive enough for a man like Gerry, he'd had his gas tank buried under the yard. Above ground there now stood an antique Texaco gas station pump, bright red and shiny, set in concrete at one side of the lane. No telling how many thousands of dollars a setup like that cost, just so the gas man, the seed corn salesmen, and I could all see it and agree with Byron: Gerry Maars was a heck of a guy.

It was cold that morning, with a stiff wind out of the north. Up on the lakes springtime was something I never saw enough of, so I'll have to say it felt good to have a little soil spread out in front of me again, straining to warm under a pale blue sky. I'll also have to say I felt a lot less poetic an hour later. The wind was still cold and I was still sitting on my fender waiting, with nothing to do but bounce pebbles off the tractor tires and admire that Texaco pump.

Over the next half hour a dozen cars sailed past in front of the house, and a school bus came back to park next to a double-wide trailer a quarter mile south. The Farm Bureau truck wound back and forth along the sections. I heard somebody's diesel roar off to the east, and shook my head. As far as I knew, even a hot dog like Gerry Maars couldn't get his crop out without using his tractors, and there they both were, sitting in front of me.

There was always the possibility, though, that I'd misunderstood. He might already be going somewhere else, and that wasn't exactly how I wanted to start things out, sitting for a couple of hours at the wrong place entirely. Gerry was a famous storyteller, and even that much would be enough to entertain the boys at coffee, dim-witted Hurd Conason's even more dim-witted son Arthur, sitting by himself out at the edge of the barnyard. I took my car out to tour the neighboring fields. There were farmers going, all right, but none of them were Gerry Maars. I drove into Keona, a couple of miles further east. The elevator was still in business, along with an old church on the main street of town. Midway Implement had a yard full of combines and a pop machine, where I bought myself a Coke.

Gerry's Fleetwood wasn't at the little cafe either, but about that time I realized how hard I was looking for the son of a gun, and turned for home. Misunderstanding or not, he could get his help somewhere else. On my way past the Bremer place there was a pickup pulled in behind the house, so I parked and walked up the lane. The four-wheel drive stood idling, and when I'd gone around the disk, I came upon Gerry Maars, his feet sticking out from under the blades.

"Hey there," he said, "you gonna work for me today?"

"I'd thought I was. What the hell time of day do you get started out here, anyway?"

"Oh, early if I can. Herschel Daley had some plans for the courthouse he thought couldn't wait. You see an inch-and-a-six-teenth out there?"

His voice had an echo to it, coming from under the machinery. As I handed the wrench down, I couldn't help thinking of my dad, who may not have been the best farmer in the world, may not have had any 4586 Internationals to pull his disk or a Fleetwood to drive back and forth to town, but who sure as heck would have had five hours of work done already by that time of day. I stood and watched Gerry's work shoes, heels kicking as he grunted and thrashed under the bank. He finally slid himself out to stand up, eyes moist under dust-feathered lashes.

"I called out here on the shop phone two or three times, when I knew I'd be late. There wasn't no answer, so I thought maybe you'd changed your mind. Thought maybe you'd decided to sleep in or something."

"There wasn't no phone that rang out here."

"Kind of hard to hear, though, I reckon, out on the highway.

While you were out driving, you see if Wendell's going yet? I seen Harry Coyne just about finished with his beans, and I ain't even got my corn in the ground yet. What time is it, almost eleven?"

"Eleven-thirty," I told him, glad to be of help.

Gerry'd always had a reputation for a temper, but after waiting around all morning I wasn't about to let him take his troubles out on me. He rattled through the drawers in his toolbox, gave up, and began rummaging through the bed of the pickup, banging aside disk blades and jacks.

"So I can't raise you, and then that goddamn Krueger boy is supposed to come out, and he ain't showed up yet either. I was laying under that disk imagining August rolling around, and me still out here planting. That's what I was picturing, August here with me still going up and down the rows by myself."

"There weren't no calls that rang out here, but you can go ahead and think that if you want. I charge the same for standing around as I do for working, so it's all the same to me."

Gerry'd found his hammer by then, and he just grunted, turning to the disk. With my piece said I finally took a second to look him over. He was still a big good-looking guy, with a solid jaw and deep-set blue eyes. The hair left around his ears had gone mostly gray, and his bright yellow polo shirt was already stained with grease where it stretched over an impressive stomach. I was never scrawny, but as he leaned in to work on the axle, his forearms bulged huge, each the size of one of my thighs.

"These kids," he said, "they tear everything up. That Shawn, the one that quit, he'd just tear things to pieces. Look at that blade over there by the truck. It's chewed half to death, and he never even noticed. Ain't that a hell of a thing? So now while the whole damn county's planting corn, I'm climbing around on the equipment. I should have run him off the first day, I'd be a hell of a lot better off."

"So what'd you figure on me doing, anyway?"

He straightened, frowning. "Like I say, first I go into town, and

Herschel Daley has to show me plans for the new furnace at the courthouse. Old Herschel don't have nothing to do anymore, so he can afford to talk. Then Irene Bremer calls, the switch on her pump's gone out. You may remember Irene, she owns this place. Poor thing's alone now, but here we are, with all these acres to plant, and I got to go play electrician. You reckon a fella like Ray Stewart is still screwing around like this? Hell no, he's been going strong for days."

Ray Stewart didn't spend all morning running around town, either, I didn't suppose, making everything in the world his business. "What did you want me to do?" I asked again.

"Oh, I reckon that'll hold, you'll have to keep an eye on it." He tossed the hammer back into the pickup and hurried around to check the oil. "You ever run one of these four-wheel drives? You've got to watch them, they've got some kind of power."

"I'll be all right."

"And that disk, it's a wide one, I got to dodge the doggone telephone poles when I move down the road, you can't hardly believe it. Equipment's a lot bigger these days, Arthur, than you might remember."

"I'll be all right, I said."

"Keep it in fourth, that way you can move right along. You want to be sure to overlap yourself. A lot of fellas don't bother, but I'm kind of particular. You may remember that about me, that I'm particular. Everybody says so. Now, you gonna be all right?"

I didn't even bother to answer, climbing past the duals to the cab. There were plenty of years I'd spent a lot more time on a tractor than I did anywhere else, but I could already tell that none of that counted. Gerry Maars was the only one who knew anything. After all, he was particular. I was just settling behind the wheel when the Krueger boy finally showed up, a pimply little hood in a denim jacket, driving a Trans Am he'd still be paying for when he retired. Gerry went over to talk to him, and when he came back the boy tagged along behind, a cigarette in one hand and a Pepsi in the other.

"Okay, Arthur, I want you to start on that field west of the ditch. Better get some fuel before you get started. You sure you're okay? You got to watch that disk, now, they're wider these days. You want me to come along with you for a round, just while you get the feel?"

Maybe I was oversensitive some, but I'd had my own dad once, and didn't much care to sit on Gerry Maars' knee while he showed me how to drive a tractor. The cab had a radio and I turned it on, hoping to drown out whatever else he might think of to yell at me.

He did yell up again, but at least he was already backing away, and I let out the clutch to swing away. The 4586 swivels in the middle to turn, but otherwise it was familiar enough; there's a secure pull and rattle to diesel that a person never really forgets. The Krueger boy had started back to the shed, with Gerry already halfway down the lane in his pickup. And that's when my right bank of disks swung head on into that damn Texaco fuel pump of his, snapping it off like a weed.

There's a sick sound of metal on metal that comes way too late to be of any help—this time long after I already had it ruined, crushed into the gravel under my blades. The Krueger boy and I watched Gerry come to a stop on the road. I turned off the radio as he drove back up the lane. He pulled around to where his antique fuel pump used to be. Then he came up alongside the cab. He pointed back to where I still had what was left of the metal housing wrapped around my disk.

"Lift 'em. Come on, lift 'em. No, the other one. The one on your left. On your left, I said. On the left! That one. Now lift it!"

My right hand offended me, shaking as I grabbed for the levers. My knees wobbled too, enough so that one foot slipped off the clutch and the tractor jerked forward, finally pulling itself free of the housing by grinding it the rest of the way into the ground. This time what the disk caught was the bumper of Gerry's pickup, swinging it around until he was looking out past an open-mouthed Billy Krueger to the shed.

As he climbed out of the truck and over the disk bank, I stayed in the cab, vaguely wishing that life could be different. Nothing grand, just that the Krueger boy hadn't been there. Or maybe that I hadn't left my car clear out by the road, where I'd have to walk all that way to get it. It's kind of pathetic, but I found myself hoping that Kendra wouldn't hear of it either, not right away. It was the last thing I wanted to do, but when I saw Gerry come up to stand below me, I opened the door.

"Damn it, Arthur, I told you them things were wide."

I looked him in the face, I couldn't very well not do that. He looked down at his crushed Texaco pump housing and back at me. I braced myself, but the next thing he did was turn to the Krueger boy.

"What the hell you lookin' at, Billy, didn't I tell you to go get seed? You see any goddamn seed around here? You see any? Where's the goddamn seed?"

He'd caught the boy by surprise, and Billy shook his head, still hypnotized by the wreckage.

"Seed don't grow in the shed, damn it, or out here in the lot, I keep telling you that. It don't grow out on the country roads, or in at the Streamliner. I need it out by the planter."

Billy didn't even have a chance to get away before Gerry'd turned to me.

"Now, Arthur, there's another fuel tank over between them bins. I didn't send you over there in the first place because it's a little on the tight side, and I don't personally like to use it. You're young, though, and like a challenge. Come on down to the field when you get finished. If Billy gets that seed out to us, we'll get some work done today after all."

Down below me Gerry kicked aside the pump housing and climbed into his pickup. He worked it until he was free of the wreckage, swung out around the disk, and without a backward glance drove away.

❈ ❈ ❈

He hadn't been kidding about that second fuel tank; it was a tight squeeze between the bins to get up close. Eventually I'd gotten the four-wheel drive maneuvered around, filled up, and took off down the road to catch him. I started off a good fifteen acres ahead, but that's not much of a margin when he wanted the ground worked twice. To top it off, before an hour had gone by, I felt myself losing the square. By the time I got to the far side of the field, I was spending most of my time in the corner, trying to finish up.

Gerry was bearing down on me by then, planting his eight rows at a pass like he was tracing with a ruler. While he finished off the end rows, I moved across the road. It was after six-thirty and getting dark by the time he pulled in behind me, shut down his own tractor, and told me to go on home. We left everything in the field, which was fine with me. As long as I didn't have to see that damn gas tank, I could almost forget it had ever been there.

I went out early the next day, figuring to get a little ways ahead. When I pulled in along the ditch, I found that a good thirty acres had already been worked. A tractor's not a race car, and in fourth gear a person averages right at six miles an hour. It took a minute to do the figures, but before long I realized that somebody'd put in another four or five hours the night before. It had to have been Gerry, who wheeled in behind me in his pickup about ten minutes later.

"My goodness, this is a pretty day. Ain't this a pretty day? Oh, this is a corn-planting day. Come on, Arthur, we got work to do."

With the headstart at least I stayed ahead, even gaining a little ground through the morning. Once Billy Krueger got there to run seed and fertilizer, Gerry started closing on me some, but I kept a safe distance between us. About four he sent the Krueger boy away.

I couldn't figure we'd be quitting quite yet, and we weren't. Billy came back with cheeseburgers from the cafe in Keona. He dropped mine off when I made my turn, and I ate them on the long pass south through the field.

By seven-thirty the sun was just settling below the horizon, red-orange from the dust that farmers had been kicking up that day all the way out to the Rockies. When I finished west of the road, I pulled up along the end rows, ready to shut it down. Instead Gerry kept the Krueger boy and had me go two miles north to sixty acres he had below the railroad. He came in behind me a couple of hours later. It was a tricky, triangle-shaped field, but even at midnight my squares still felt true. A half hour later I wasn't sure whether to trust my eyes or not, when I saw Gerry flashing his lights to signal me, and knew we were finally done for the day.

Times had changed. I remember days when my dad would come in from planting late, back when Byron and I were boys. We thought it so far out of the ordinary, imagine, all of us eating supper at ten o'clock at night! Apparently International Harvester had built their headlights to go a little later than that, and I soon found out that there were very few of the nighttime hours that Gerry didn't end up using. The third day we went past midnight again, the day after that until almost one. We finished Skidmore's a little after two the night of the fifth day. The sixth and seventh days, when the Krueger boy didn't come in, we finally realized he'd quit.

I imagine Gerry'd thought he'd run me off too, eventually, but Byron and I had been raised poor, not lazy. Once I got his schedule figured, I'd always be out there a good hour before he was, and I never came calling to leave. On breakdowns I was a pretty good hand to have, too, once I got over my surprise that they could even happen to the great Gerry Maars. All I had to get back was my patience. Sometimes a bearing would break in the most awkward place imaginable, and we'd lay out there half the afternoon, trying to work the damn thing loose. Finally Gerry'd climb up into the cab to get his glasses, and I'd run into Midway before they closed. I'd leave him there tinkering and tapping, pounding, cussing it, and nine times out of ten, by the time I got back, he had it off. We'd get the thing put back together, dig the dust out of our eyes, and make up lost time until dawn.

You've got to figure breakdowns are part of the business, especially when you're bouncing over all that ground. What I couldn't figure, not with the acreage he was carrying, were all the distractions that seemed to come with being important. It wasn't enough that he had breakfast in town; I'd watch from my tractor as a steady stream of visitors made their way out to the field. It didn't matter where we were; one of the other council members, somebody from the tax assessor's office, the deputy auditor—any one of them might be waving to him at the end of the field. Every second morning or so Irene Bremer would have something that demanded his attention too. If she missed him at home, she'd call out to the elevator. Everybody at the elevator knew her too, but they were too busy playing grab-ass to look in on her themselves, so the fertilizer truck driver would bring the message out to Gerry. Many were the days when I'd see his tractor idling at one end of the field, and the pickup truck gone. He'd be back a couple hours later, all worked up about the time he'd lost, and in more of a hurry than ever.

That meant I got a chance to get a little ways ahead, and whenever I did, I'd come back and help him pull seed. It was easy enough to understand why the Krueger boy quit: I could still remember how it felt, standing out there all sweaty and covered with dirt while the other kids drove by on their way to town. In my case I'd already *been* to town, so nighttime out in the wide empty fields of spring suited me fine. We finally came into Smiley's the third week I was there.

As far as property goes, the Smiley acreage was nothing special, poorly drained and weedy like a lot of that ground below the tracks. To my family, though, it had been about all there was to our little farm. That land we rented from Smiley was where we raised alfalfa and wheat, drove our hogs over in December to glean the corn. It was where Byron and I had kept our heads down for what seemed years at a time, eating dust all spring on the tractor and freezing half to death out with the cornpicker in the fall.

Now, with an operation like Gerry's, things went a hell of a lot easier. The fences were gone, which made for better maneuvering. We had equipment that was big enough to do the job. After a dozen cans of Freon we'd even got the air-conditioning fixed in the big four-wheel drive, so the cab was cool under all that glass. There was a high spot that I still remembered north of the ditch bank— going west on that big tractor, I could see the roof of Byron's ranch house, and beyond it the battered roof of the barn.

I don't know what mission of mercy Gerry was off on while I disked that first morning at Smiley's, but when he got back about two o'clock, he'd brought hamburgers with him. We sat back on the tailgate to eat, under sunshine and a gentle May breeze. Eventually I gathered up my trash, and when I went over to check my banks, Gerry followed.

"You know, Arthur," he said, "I remember your dad out here. I guess you probably do too."

I told him I remembered all of us, we'd spent a lot of time in that field.

"Yeah, I sure do remember your dad. Back when I was first starting out, I remember one time seeing him. It was in the spring like this, over at the Bremer place, the first year I was here. One morning I was getting my planter rigged up, and your dad came wheeling up into the field. He had that old white Dodge pickup, maybe you remember. Anyway, he came charging over like an old bull. Hell, your dad *was* a bull back then, built like a tree stump. And he said, 'Just what are you getting ready to do there, Mr. Maars?'"

Gerry laughed, remembering.

"I was a smart aleck back then, full of sass. So I come back and say, 'I don't guess maybe you've ever seen one of these before, Hurd. It's a corn planter. I'm fixing to plant myself some corn.'

"He wasn't much of a kidder, your dad. A hell of a good man, but kind of grim."

"'The hell you are,' he said.

"'The hell I ain't,' I said. Oh, I was a hothead back then. See, he was an established member of the community and all, and there I was just a young squirt, about your age. So he thought he'd just tell me how things were gonna be.

"'The hell you are,' he said again. 'The hell I ain't.' We went round and round. And it turned out it was the corn bores, see, they didn't want nobody planting until the twentieth of May. That way you broke the cycle. Ain't that something? They had it all figured out. You wait until after the twentieth. And now the whole damn county is laughing at us when they drive by and see us still on corn. Ray Stewart, Lindell, they're all proud of themselves 'cause they're damn near finished. Back then they'd just be getting started!"

"So what'd you do?"

"Do when?"

"Back then, how'd it end up? What'd you end up doing?"

"It ended up I didn't plant, I guess, I can't remember for sure. I must have had to wait, you know, that's just the way it was. You had to break the cycle. 'Course, I was glad when them chemicals came along, we all were, it made things a little easier. I can't hardly see us waiting around all spring anymore, with the kind of acreage we got now."

Gerry wasn't one to wait around when lunch was over either, but there at Smiley's I felt him still behind me while I chimed my way along the disk blades, making sure they were tight. I climbed up to throw the hammer behind the seat, and leaned out to say so long. Gerry still hadn't moved from in front of my bank. I didn't guess he would want me to disk him under with the weeds, so I climbed back down.

"See," he said, "people acted together back then. They were more of a community. That's the way life was."

"It was different back then, that's for sure."

"That's exactly right. You got what I'm saying. Things were different. You're following me."

I told him I thought I was.

"See, Arthur, I know you are. But you tell that same story to Shawn, or Billy Krueger, and they just look at you like you're speaking Chinese. They just want to drink beer and smoke that pot all night, and then come out and hot-rod around on the tractors. They don't see that people put their whole lives into these places, working things out. They don't see that."

"Well," I said, "we ain't gonna work nothing out standing around here."

Whatever it was he was driving at, Gerry still wasn't satisfied. He stood up next to me in that empty field at Smiley's, rubbing his big knuckles against his forearm as he looked out over the prairie.

"Arthur, I didn't take this place away from your dad. You know that, don't you? You know I wouldn't do that?"

"It doesn't matter," I said.

"It does to me. Gene Honnager was the one who came running down here one year when money was cheap, rented this place right out from under him. That's what happened. I know that made it hard on Hurd. I know it did. But a lot of people don't give a damn anymore. Greed drives the engine."

I couldn't help it, I laughed.

Gerry managed to look hurt. "That's the way it was, Arthur. And even then Gene couldn't work it worth a damn, he didn't have the patience. He just run on to where the money came easier. When Gene let it go, old Smiley's widow called me up, begging me to take it off her hands. So I did. That's why we're here. Your dad was already sick by then."

"It sure doesn't matter now."

"It does to me, I said, damn it, I ain't no Gene Honnager. That ain't the way I do my business. There was a time here in Haskell County when we wouldn't even have allowed what Gene done.

Somebody would have kicked his ass for him. People acted together back then."

I could have told him it didn't matter again, but Gerry was back looking out over the prairie. I couldn't think of anything more foolish anyway, than for us to be standing out there in a half-planted field while he apologized for something he hadn't done. For taking on land somebody else once had rented. In modern times it didn't make much sense, and Gerry must have realized it too, about then, because he went over and climbed into his cab. He pulled past the end rows, dropped his marker, and took off to the east, sowing corn. That was the last I saw of him until after midnight. I was on my last pass back along the fence when I saw his headlights blinking, and knew that we were at the end of another day.

Chapter 10

Owning It All

William Kittredge

Probably in no region of America did settlers exploit the land with more untempered zeal than in the West. Mountains were opened and crumbled to yield silver and gold; rivers were run dry to grow grain; animals were slaughtered in response to whims as often as needs. Yet so vast and, in many places, so resilient was the land that decades and even generations passed before the ill effects became manifest. Agrarian sentiments never disappeared, but they stood little chance against the powerful frontier myths of miners and cattlemen.

Growing up on and later managing a ranch in southeastern Oregon, writer William Kittredge absorbed the stories and land-use values of this exploitive economy. Only slowly did he sense the many forms of poverty—moral, social, and ecological—that were its inevitable fruits. In this frank memoir, Kittredge recounts his growing sense of malaise and then betrayal as he came to recognize that he and his kind "had wrecked all we had not left untouched." The root cause, as he came to see it, was the urge to "own it all." Once gripped by that addiction, it is hard for an individual, much less a culture, to break free.

Like Mendelson's history and Kurtz's fiction, Kittredge's memoir stands at the center of the agrarian critique of modern American culture.

Imagine the slow history of our country in the far reaches of southeastern Oregon, a backlands enclave even in the American West, the first settlers not arriving until a decade after the end of the Civil War. I've learned to think of myself as having had the luck to grow up at the tail end of a way of existing in which people lived in every-day proximity to animals on territory they knew more precisely than the patterns in the palms of their hands.

In Warner Valley we understood our property as others know their cities, a landscape of neighborhoods, some sacred, some demonic, some habitable, some not, which is as the sea, they tell me, is understood by fishermen. It was only later, in college, that I learned it was possible to understand Warner as a fertile oasis in a vast featureless sagebrush desert.

Over in that other world on the edge of rain-forests which is the Willamette Valley of Oregon, I'd gone to school in General Agriculture, absorbed in a double-bind sort of learning, studying to center myself in the County Agent/Corps of Engineers mentality they taught and at the same time taking classes from Bernard Malamud and wondering with great romantic fervor if it was in me to write the true history of the place where I had always lived.

Straight from college I went to Photo Intelligence work in the Air Force. The last couple of those years were spent deep in jungle on the island of Guam, where we lived in a little compound of cleared land, in a quonset hut.

The years on Guam were basically happy and bookish: we were newly married, with children. A hundred or so yards north of our quonset hut, along a trail through the luxuriant undergrowth between coconut palms and banana trees, a ragged cliff of red porous volcanic rock fell directly to the ocean. When the Pacific typhoons came roaring in, our hut was washed with blowing spray from the great breakers. On calm days we would stand on the cliff at that absolute edge of our jungle and island, and gaze out across to the island of Rota, and to the endlessness of ocean beyond, and I would marvel at my life, so far from southeastern Oregon.

And then in the late fall of 1958, after I had been gone from Warner Valley for eight years, I came back to participate in our agriculture. The road in had been paved, we had Bonneville Power on lines from the Columbia River, and high atop the western rim of the valley there was a TV translator, which beamed fluttering pictures from New York and Los Angeles direct to us.

And I had changed, or thought I had, for a while. No more daydreams about writing the true history. Try to understand my excitement as I climbed to the rim behind our house and stood there by our community TV translator. The valley where I had always seen myself living was open before me like another map and playground, and this time I was an adult, and high up in the War Department. Looking down maybe 3,000 feet into Warner, and across to the high basin and range desert where we summered our cattle, I saw the beginnings of my real life as an agricultural manager. The flow of watercourses in the valley was spread before me like a map, and I saw it as a surgeon might see the flow of blood across a chart of anatomy, and saw myself helping to turn the fertile homeplace of my childhood into a machine for agriculture whose features could be delineated with the same surgeon's precision in my mind.

It was work which can be thought of as craftsmanlike, both artistic and mechanical, creating order according to an ideal of beauty based on efficiency, manipulating the forces of water and soil, season and seed, manpower and equipment, laying out functional patterns for irrigation and cultivation on the surface of our valley. We drained and leveled, ditched and pumped, and for a long while our crops were all any of us could have asked. There were over 5,000 water control devices. We constructed a perfect agricultural place, and it was sacred, so it seemed.

❖ ❖ ❖

Agriculture is often envisioned as an art, and it can be. Of course there is always survival, and bank notes, and all that. But your basic bottom line on the farm is again and again some notion of how life should be lived. The majority of agricultural people, if you press them hard enough, even though most of them despise sentimental abstractions, will admit they are trying to create a good place, and to live as part of that goodness, in the kind of connection which with fine reason we call *rootedness*. It's just that there is good art and bad art.

These are thoughts which come back when I visit eastern Oregon. I park and stand looking down into the lava-rock and juniper-tree canyon where Deep Creek cuts its way out of the Warner Mountains, and the great turkey buzzard soars high in the yellow-orange light above the evening. The fishing water is low, as it always is in late August, unfurling itself around dark and broken boulders. The trout, I know, are hanging where the currents swirl across themselves, waiting for the one entirely precise and lucky cast, the Renegade fly bobbing toward them.

Even now I can see it, each turn of water along miles of that creek. Walk some stretch enough times with a fly rod and its configurations will imprint themselves on your being with Newtonian exactitude. Which is beyond doubt one of the attractions of such fishing—the hours of learning, and then the intimacy with a living system that carries you beyond the sadness of mere gaming for sport.

What I liked to do, back in the old days, was pack in some spuds and an onion and corn flour and spices mixed up in a plastic bag, a small cast-iron frying pan in my wicker creel and, in the late twilight on a gravel bar by the water, cook up a couple of rainbows over a fire of snapping dead willow and sage, eating alone while the birds flitted through the last hatch, wiping my greasy fingers on my pants while the heavy trout began rolling at the lower ends of the pools.

The canyon would be shadowed under the moon when I walked out to show up home empty-handed, to sit with my wife over a drink of whiskey at the kitchen table. Those nights I would go to bed and sleep without dreams, a grown-up man secure in the house and the western valley where he had been a child, enclosed in a topography of spirit he assumed he knew more closely than his own features in the shaving mirror.

So, I ask myself, if it was such a pretty life, why didn't I stay? The peat soil in Warner Valley was deep and rich, we ran good cattle, and my most sacred memories are centered there. What could run me off?

Well, for openers, it got harder and harder to get out of bed in the mornings and face the days, for reasons I didn't understand. More and more I sought the comfort of fishing that knowable creek. Or in winter the blindness of television.

My father grew up on a homestead place on the sagebrush flats outside Silver Lake, Oregon. He tells of hiding under the bed with his sisters when strangers came to the gate. He grew up, as we all did in that country and era, believing that the one sure defense against the world was property. I was born in 1932, and recall a life before the end of World War II in which it was possible for a child to imagine that his family owned the world.

Warner Valley was largely swampland when my grandfather bought the MC Ranch with no down payment in 1936, right at the heart of the Great Depression. The outside work was done mostly by men and horses and mules, and our ranch valley was filled with life. In 1937 my father bought his first track-layer, a secondhand RD6 Caterpillar he used to build a 17-mile diversion canal to carry the spring floodwater around the east side of the valley, and we were on our way to draining all swamps. The next year he bought an RD7 and a John Deere 36 combine which cut an 18-foot swath, and we were deeper into the dream of power over nature and men, which I had begun to inhabit while playing those long-ago games of war.

The peat ground left by the decaying remnants of ancient tule beds was diked into huge undulating grainfields—Houston Swamp with 750 irrigated acres, Dodson Lake with 800—a final total of almost 8,000 acres under cultivation, and for reasons of what seemed like common sense and efficiency, the work became industrialized. Our artistry worked toward a model whose central image was the machine.

The natural patterns of drainage were squared into dragline ditches, the tules and the aftermath of the oat and barley crops were burned—along with a little more of the combustible peat soil every year. We flood-irrigated when the water came in spring, drained in late March, and planted in a 24-hour-a-day frenzy which began around April 25 and ended—with luck—by the 10th of May, just as leaves on the Lombardy poplar were breaking from their buds. We summered our cattle on more than a million acres of Taylor Grazing Land across the high lava rock and sagebrush desert out east of the valley, miles of territory where we owned most of what water there was, and it was ours. We owned it all, or so we felt. The government was as distant as news on the radio.

The most intricate part of my job was called "balancing water," a night and day process of opening and closing pipes and redwood headgates and running the 18-inch drainage pumps. That system was the finest plaything I ever had.

And despite the mud and endless hours, the work remained play for a long time, the making of a thing both functional and elegant. We were doing God's labor and creating a good place on earth, living the pastoral yeoman dream—that's how our mythology defined it, although nobody would ever have thought to talk about work in that way.

And then it all went dead, over years, but swiftly.

You can imagine our surprise and despair, our sense of having been profoundly cheated. It took us a long while to realize some unnamable thing was wrong, and then we blamed it on ourselves, our inability to manage enough. But the fault wasn't ours, beyond the fact that we had all been educated to believe in a grand bad factory-land notion as our prime model of excellence.

We felt enormously betrayed. For so many years, through endless efforts, we had proceeded in good faith, and it turned out we had wrecked all we had not left untouched. The beloved migratory rafts of waterbirds, the green-headed mallards and the redheads and canvasbacks, the cinnamon teal and the great Canadian Honkers, were mostly gone along with their swampland habitat. The hunting, in so many ways, was no longer what it had been.

We wanted to build a reservoir, and litigation started. Our laws were being used against us, by people who wanted a share of what we thought of as our water. We could not endure the boredom of our mechanical work, and couldn't hire anyone who cared enough to do it right. We baited the coyotes with 1080, and rodents destroyed our alfalfa; we sprayed weeds and insects with 2-4-D Ethyl and Malathion, and Parathion for clover mite, and we shortened our own lives.

In quite an actual way we had come to victory in the artistry of our playground warfare against all that was naturally alive in our native home. We had reinvented our valley according to the most persuasive ideal given us by our culture, and we ended with a landscape organized like a machine for growing crops and fattening cattle, a machine that creaked a little louder each year, a dreamland gone wrong.

One of my strongest memories comes from a morning when I was maybe 10 years old, out on the lawn before our country home in spring, beneath a bluebird sky. I was watching the waterbirds coming off the valley swamps and grainfields where they had been feeding overnight. They were going north to nesting grounds on the Canadian tundra, and that piece of morning, inhabited by the sounds of their wings and their calling in the clean air, was wonderfilled and magical. I was enclosed in a living place.

No doubt that memory has persisted because it was a sight of possibility which I will always cherish—an image of the great good place rubbed smooth over the years like a river stone, which I touch again as I consider why life in Warner Valley went so seriously haywire. But never again in my lifetime will it be possible for a child to stand out on a bright spring morning in Warner Valley and watch the waterbirds come through in enormous, rafting vee-shaped flocks of thousands—and I grieve.

My father is a very old man. A while back we were driving up the Bitterroot Valley of Montana, and he was gazing away to the mountains. "They'll never see it the way we did," he said, and I wonder what he saw.

We shaped our piece of the West according to the model provided by our mythology, and instead of a great good place such order had given us enormous power over nature, and a blank perfection of fields.

❧   ❧   ❧

A mythology can be understood as a story that contains a set of implicit instructions from a society to its members, telling them what is valuable and how to conduct themselves if they are to preserve the things they value.

The teaching mythology we grew up with in the American West is a pastoral story of agricultural ownership. The story begins with a vast innocent continent, natural and almost magically alive, capable of inspiring us to reverence and awe, and yet savage, a wilderness. A good rural people come from the East, and they take the land from its native inhabitants, and tame it for agricultural purposes, bringing civilization: a notion of how to live embodied in law. The story is as old as invading armies, and at heart it is a racist, sexist, imperialist mythology of conquest; a rationale for violence—against other people and against nature.

At the same time, that mythology is a lens through which we continue to see ourselves. Many of us like to imagine ourselves as honest yeomen who sweat and work in the woods or the mines or the fields for a living. And many of us are. We live in a real family, a work-centered society, and we like to see ourselves as people with the good luck and sense to live in a place where some vestige of the natural world still exists in working order. Many of us hold that natural world as sacred to some degree, just as it is in our myth. Lately, more and more of us are coming to understand our society in the American West as an exploited colony, threatened by greedy outsiders who want to take our sacred place away from us, or at least to strip and degrade it.

In short, we see ourselves as a society of mostly decent people who live with some connection to a holy wilderness, threatened by those who lust for power and property. We look for Shane to come riding out of the Tetons, and instead we see Exxon and the Sierra Club. One looks virtually as alien as the other.

And our mythology tells us we own the West, absolutely and morally—we own it because of our history. Our people brought law to this difficult place, they suffered and they shed blood and they survived, and they earned this land for us. Our efforts have surely earned us the right to absolute control over the thing we created. The myth tells us this place is ours, and will always be ours, to do with as we see fit.

That's a most troubling and enduring message, because we want to believe it, and we do believe it, so many of us, despite its implicit ironies and wrongheadedness, despite the fact that we took the land from someone else. We try to ignore a genocidal history of violence against the Native Americans.

In the American West we are struggling to revise our dominant mythology, and to find a new story to inhabit. Laws control our lives, and they are designed to preserve a model of society based on values learned from mythology. Only after re-imagining our myths can we coherently remodel our laws, and hope to keep our society in a realistic relationship to what is actual.

In Warner Valley we thought we were living the right lives, creating a great precise perfection of fields, and we found the mythology had been telling us an enormous lie. The world had proven too complex, or the myth too simpleminded. And we were mortally angered.

The truth is, we never owned all the land and water. We don't even own very much of them, privately. And we don't own anything absolutely or forever. As our society grows more and more complex and interwoven, our entitlement becomes less and less absolute, more and more likely to be legally diminished. Our rights to property will never take precedence over the needs of society. Nor should they, we all must agree in our grudging hearts. Ownership of property has always been a privilege granted by society, and revokable.

❊   ❊   ❊

Down by the slaughterhouse my grandfather used to keep a chicken-wire cage for trapping magpies. The cage was as high as a man's head, and mounted on a sled so it could be towed off and cleaned. It worked on the same principle as a lobster trap. Those iridescent black-and-white birds could get in to feed on the intestines of butchered cows—we never butchered a fat heifer or steer for our own consumption, only aged dry cows culled from the breeding herd—but they couldn't get out.

Trapped under the noontime sun, the magpies would flutter around in futile exploration for a while, and then would give in to a great sullen presentiment of their fate, just hopping around picking at leftovers and waiting.

My grandfather was Scots-English, and a very old man by then, but his blue eyes never turned watery and lost. He was one of those cowmen we don't see so often anymore, heedless of most everything outside his playground, which was livestock and seasons and property, and, as the seasons turned, more livestock and more property, a game which could be called accumulation.

All the notes were paid off, and you would have thought my grandfather would have been secure, and released to ease back in wisdom.

But no such luck. It seemed he had to keep proving his ownership. This took various forms, like endless litigation, which I have heard described as the sport of kings, but the manifestation I recall most vividly was that of killing magpies.

In the summer the ranch hands would butcher in the after-supper cool of an evening a couple of times a week. About once a week, when a number of magpies had gathered in the trap, maybe 10 or 15, my grandfather would get out his lifetime 12-gauge shot-gun and have someone drive him down to the slaughterhouse in his dusty, ancient gray Cadillac, so he could look over his catch and get down to the business at hand. Once there, the ritual was slow and dignified, and always inevitable as one shoe after another.

The old man would sit there a while in his Cadillac and gaze at the magpies with his merciless blue eyes, and the birds would stare back with their hard black eyes. The summer dust would settle around the Cadillac, and the silent confrontation would continue. It would last several minutes.

Then my grandfather would sigh, and swing open the door on his side of the Cadillac, and climb out, dragging his shotgun behind him, the pockets of his gray gabardine suit-coat like a frayed uni-form bulging with shells. The stock of the shotgun had been bro-ken sometime deep in the past, and it was wrapped with fine brass wire, which shone golden in the sunlight while the old man thumbed shells into the magazine. All this without saying a word.

In the ear of my mind I try to imagine the radio playing softly in the Cadillac, something like "Room Full of Roses" or "Candy Kisses," but there was no radio. There was just the ongoing hum of insects and the clacking of the mechanism as the old man pumped a shell into the firing chamber.

He would lift the shotgun, and from no more than 12 feet, sighting down that barrel where the bluing was mostly worn off, through the chicken wire into the eyes of those trapped magpies, he would kill them one by one, taking his time, maybe so as to prove that this was no accident.

He would fire and there would be a minor explosion of blood and feathers, the huge booming of the shotgun echoing through the flattening light of early afternoon, off the sage-covered hills and down across the hay meadows and the sloughs lined with dagger-leafed willow, frightening great flights of blackbirds from the fence lines nearby, to rise in flocks and wheel and be gone.

"Bastards," my grandfather would mutter, and then he would take his time about killing another, and finally he would be finished and turn without looking back, and climb into his side of the Cadillac, where the door still stood open. Whoever it was whose turn it was that day would drive him back up the willow-lined lane through the meadows to the ranch house beneath the Lombardy poplar, to the cool shaded living room with its faded linoleum where the old man would finish out his day playing pinochle with my grandmother and anyone else he could gather, sometimes taking a break to retune a favorite program on the Zenith Trans-Oceanic radio.

No one in our family, so far as I ever heard, knew any reason why the old man had come to hate magpies with such specific intensity in his old age. The blackbirds were endlessly worse, the way they would mass together in flocks of literally thousands, to strip and thrash in his oat and barley fields, and then feed all fall in the bins of grain stockpiled to fatten his cattle.

"Where is the difference?" I asked him once, about the magpies.

"Because they're mine," he said. I never did know exactly what he was talking about, the remnants of entrails left over from the butchering of culled stocker cows, or the magpies. But it became clear he was asserting his absolute lordship over both, and over me, too, so long as I was living on his property. For all his life and most of mine the notion of property as absolute seemed like law, even when it never was.

Most of us who grew up owning land in the West believed that any impairment of our right to absolute control of that property was a taking, forbidden by the so-called "taking clause" of the Constitution. We believed regulation of our property rights could never legally reduce the value of our property. After all, what was the point of ownership if it was not profitable? Any infringement on the control of private property was a communist perversion.

But all over the West, as in all of America, the old folkway of property as an absolute right is dying. Our mythology doesn't work anymore.

We find ourselves weathering a rough winter of discontent, snared in the uncertainties of a transitional time and urgently yearning to inhabit a story that might bring sensible order to our lives—even as we know such a story can only evolve through an almost literally infinite series of recognitions of what, individually, we hold sacred. The liberties our people came seeking are more and more constrained, and here in the West, as everywhere, we hate it.

Simple as that. And we have to live with it. There is no more running away to territory. This is it, for most of us. We have no choice but to live in community. If we're lucky we may discover a story that teaches us to abhor our old romance with conquest and possession.

My grandfather died in 1958, toppling out of his chair at the pinochle table, soon after I came back to Warner, but his vision dominated our lives until we sold the ranch in 1967. An ideal of absolute ownership that defines family as property is the perfect device for driving people away from one another. There was a rule in our family. "What's good for the property is good for you."

"Every time there was more money we bought land," my grandmother proclaimed after learning my grandfather had been elected to the Cowboy Hall of Fame. I don't know if she spoke with pride or bitterness, but I do know that, having learned to understand love as property, we were all absolutely divided at the end; relieved to escape amid a litany of divorce and settlements, our family broken in the getaway.

I cannot grieve for my grandfather. It is hard to imagine, these days, that any man could ever again think he owns the birds.

❧ ❧ ❧

Thank the Lord there were other old men involved in my upbringing. My grandfather on my mother's side ran away from a Germanic farmstead in Wisconsin the year he was fourteen, around 1900, and made his way to Butte. "I was lucky," he would say. "I was too young to go down in the mines, so they put me to sharpening steel."

Seems to me such a boy must have been lucky to find work at all, wandering the teeming difficult streets of the most urban city in the American West. "Well, no," he said. "They put you to work. It wasn't like that. They were good to me in Butte. They taught me a trade. That's all I did was work. But it didn't hurt me any."

After most of ten years on the hill—broke and on strike, still a very young man—he rode the rails south to the silver mines in what he called "Old Mexico," and then worked his way back north through the mining country of Nevada in time for the glory days in Goldfield and Rhyolite and Tonopah. At least those are the stories he would tell. "This Las Vegas," he would say. "When I was there you could have bought it all for a hundred and fifty dollars. Cost you ten cents for a drink of water."

To my everlasting sadness, I never really quizzed him on the facts. Now I look at old photographs of those mining camps, and wonder. It's difficult for me to imagine the good gentle man I knew walking those tough dusty streets. He belonged, at least in those Butte days, to the International Brotherhood of Blacksmiths and Helpers. I still have his first dues card. He was initiated July 11, 1904, and most of the months of 1904 and 1905 are stamped, DUES PAID.

❧ ❧ ❧

Al died in an old folks' home in Eugene, Oregon. During the days of his last summer, when he knew the jig was up, a fact he seemed to regard with infallible good humor, we would sit in his room and listen to the aged bemused woman across the hall chant her litany of childhood, telling herself that she was somebody and still real.

It was always precisely the same story, word by particular word. I wondered then how much of it was actual, lifting from some deep archive in her memory, and now I wonder how much of it was pure sweet invention, occasioned by the act of storytelling and by the generative, associative power of language. I cannot help but think of ancient fires, light flickering on the faces of children and story-tellers detailing the history of their place in the scheme of earth.

The story itself started with a screen-door slamming and her mother yelling at her when she was a child coming out from the back porch of a white house, and rotting apples on the ground under the trees in the orchard, and a dog which snapped at the flies. "Mother," she would exclaim in exasperation, "I'm fine."

The telling took about three minutes, and she told it like a story for grandchildren. "That's nice," she would say to her dog. "That's nice."

Then she would lapse into quiet, rewinding herself, seeing an old time when the world contained solace enough to seem complete, and she would start over again, going on until she had lulled herself back into sleep. I would wonder if she was dreaming about that dog amid the fallen apples, snapping at flies and yellowjackets.

At the end she would call the name of that dog over and over in a quavering, beseeching voice—and my grandfather would look to me from his bed and his eyes would be gleaming with laughter, such an old man laughing painfully, his shoulders shaking, and wheezing.

"Son of a bitch," he would whisper, when she was done calling the dog again, and he would wipe the tears from his face with the sleeve of his hospital gown. *Son of a bitch.* He would look to me again, and other than aimless grinning acknowledgment that some mysterious thing was truly funny, I wouldn't know what to do, and then he would look away to the open window, beyond which a far-off lawn mower droned, like this time he was the one who was embarrassed. Not long after that he was dead, and so was the old woman across the hall.

"Son of a bitch," I thought, when we were burying Al one bright afternoon in Eugene, and I found myself suppressing laughter. Maybe it was just a way of ditching my grief for myself, who did not know him well enough to really understand what he thought was funny. I have Al's picture framed on my wall, and I can still look to him and find relief from the old insistent force of my desire to own things. His laughter is like a gift.

Chapter 11

The Wealth of Nature

Donald Worster

*The United States thrived economically in the nineteenth and twentieth
centuries largely because it had unleashed so fully the entrepreneurial
energies of its people. The landscape was a vast one, and traditional
social constraints, in place back in Europe, often withered and even disappeared. It was a country with plenty of room and little sense of history;
it was a culture of arrested adolescence, propelled by the yearning, as
William Kittredge so aptly put it, to own it all.*

*If that lack of history and historical awareness has aided enterprise, it
has stood as an obstacle to the emergence of a more settled, rooted culture. It has been hard for Americans to see clearly where they have been,
to spot the cultural errors they have made and identify the paths they
have chosen not to take.*

*Few historians have worked harder to clarify these truths than Don-
ald Worster, who teaches in his native state at the University of Kansas.
In the following selection, adapted from* The Wealth of Nature: Environmental History and the Ecological Imagination, *Worster probes
the cultural origins of our environmental predicament. He finds them
largely in the seventeenth and eighteenth centuries, the age when a new
worldview gained power, a view that Worster terms materialism. In its
economic and scientific forms, materialism purged nature of mystery and
inherent value; it embraced a progressive, linear view of change; and it
exalted human desires and maximum economic productivity. Once mysterious and God-crafted, nature became mere inert matter, valuable only
insofar as it was useful to humans and available for humans to manipulate at will. Worster identifies the need for a new worldview, a more*

respectful, virtuous one that admits our "real and indispensable" depend-
ence on the greater "economy of nature." Worster's title plays on that of
the classic work of Adam Smith, The Wealth of Nations—*the preem-*
inent summation, Worster claims, of the materialistic persuasion.

Whoever made the dollar bill green had a right instinct. There is a connection, profound and yet so easy to ignore, between the money in our pocket and the green earth, though the connection is more than color. The dollar bill needs paper, which is to say it needs trees, just as our wealth in general derives from nature, from the forest, the earth and waters, the soil. That these are all limited and finite is easy to see, and so also must be wealth; it can never be unlimited, though it can be expanded and multiplied by human ingenuity. Somewhere on the dollar bill that message might be printed, a warning that you hold in your hand a piece of the limited earth that should be handled with respect: "In God we trust; on nature we must depend."

The public is beginning to understand that connection in at least a rudimentary way and to realize that taking better care of the earth will cost money, will lower the standard of living as it is conventionally defined, and will interfere with freedom of enterprise. By the evidence of opinion polls, something like three out of four Americans say they are ready to accept those costs, a remarkable development in our history. The same can be said for almost every other nation on earth, even the poorest, who are learning that, in their own long-term self-interest, the preservation of nature is a cost they ought to pay, though they may demand that the rich nations assume some of the cost. Having money in one's pocket, no matter how green its color, is no longer the unexamined good it once was. Many have come to realize that wealth might be a kind of poverty.

The human species, according to a team of Stanford biologists, is now consuming or destroying 40 percent of the net primary terrestrial production of the planet: that is, nearly one-half of all the energy fixed by photosynthesis on the land. We are harvesting it, drastically reorganizing it, or losing it through urbanization and desertification in order to support our growing numbers and even faster growing demands. In addition, much of the remaining 60 percent is profoundly affected by the pumping and burning of fossil fuels, the spreading about of so many chemicals new to evolution, the accelerating interventions into the water cycle, the atmosphere, the climate.

That impact is sure to increase, as more than ever we seek to turn the earth into wealth. The United Nations now projects that world population will grow to more than eight billion by 2025, then go on to ten billion before it stabilizes toward the end of the next century, so that the present heavy impact on ecosystems has only just begun. One-fourth of the world's total stock of plant and animal species are at risk of being eliminated in the next twenty years. About half of the rain forests in tropical areas have already been lost to deforestation, and an additional area the size of Kansas is being lost every year to clear-cutting for timber, cattle grazing, and other uses. The increased burning of fossil fuels is beginning to raise atmospheric carbon dioxide levels so rapidly that global shifts in climate appear imminent. Those are some of the costs forcing the public to re-evaluate the ends and means of wealth.

Suddenly, we humans are waking up to the massive influence we are having on the planet in the pursuit of greater production and are beginning to wonder whether the wealth is excessive and how long it can last. We are beginning to fear that we cannot really manage this enormous productive apparatus that we have superimposed on nature. The earth has begun to look like a savings and loan office six months after bankruptcy: the furniture disappearing, the water cooler empty, the looks on the faces of the office staff blank or bewildered. . . .

✽ ✽ ✽

The environmental crisis that has emerged over the last half century, though unprecedented in scope and complexity, is not the first in history. The human past reveals a long chain of crises stemming from a lack of knowledge or foresight, though typically before the modern era they were highly localized. The migrants from Asia, for instance, who entered North America some 30,000 to 40,000 years ago had no idea, as they stalked and slaughtered the hairy mammoths gathering around a waterhole, that they would one day run out of easy meat and then would have to make drastic changes in their weapons and hunting targets. I am sure too that the ancient Mesopotamians never imagined, as they dug their irrigation ditches to raise crops in the desert, that one day they would find those ditches filling with silt and their fields poisoned with salt. Much of human history appears as a succession of ecological surprises, many of them tragic, that communities have encountered on their way to dinner or a warm bed.

From our own vantage today it might seem that all those past people of history failed to achieve some enduring method of getting a living from the earth because they were ignorant of how the natural world works. Had the Pleistocene hunters had a few of our computer-armed population biologists advising them, had the Mesopotamians had the advantage of modern hydraulic engineering, no surprises would have happened. Those folks lived in illiterate, irrational times, in contrast to our state of enlightenment.

But if all that was lacking in the past was scientific understanding, then we men and women of the late twentieth century surely ought to be beyond almost all possibility of ecological surprise and failure. Somebody has calculated that one out of every two scientists who ever lived is alive today. We ought, therefore, to have enough of them around, and enough laboratories and research programs, to manage our relations supremely well with the natural world, so well that we could leave all fear of failure behind us. This ought to be the age of absolutely reliable control, when human life runs along in a steady course, when the earth hums like a Japanese factory, when no one ever sweats and no one ever has to worry about their children getting skin cancer.

Perhaps the single most impressive lesson of history, however, is that, despite all our scientific expertise, all our investment in productive machinery, all the wealth we have acquired, we still have not escaped from the inadequacy of our knowledge. On the contrary, each year we encounter greater ecological steering problems than before, which we are unprepared to handle. And this managerial crisis threatens to go on increasing in seriousness well into the next century.

In 1967, when the phrase "environmental" or "ecological crisis" first began to appear widely in the press, the distinguished medievalist Lynn White, Jr., presented a historical analysis of our predicament that deserves to be read and reread regularly by the world's policy makers, though I will argue in a moment that it was ultimately an unpersuasive analysis. White doubted that we could resolve the crisis by "applying to our problems more science and more technology."[1] In fact, trying to resolve it without understanding its roots, as our technicians seemed to be doing, ran the risk of making it even worse. He was not advocating that we do nothing unless we can do something grandiose, nor was he unsympathetic toward the technicians pressured to find some immediate, pragmatic solutions. But as a historian he saw in the crisis some larger cultural challenges that too often scientists, engineers, economists, politicians, and others had not even studied, let alone understood, and he insisted that addressing those larger challenges must be part of any lasting resolution of the crisis.

"Human ecology," White pointed out, "is deeply conditioned by beliefs about our nature and destiny—that is, by religion." He argued that the environmental crisis emerged, not just yesterday, but over the long sweep of Western civilization. Specifically, it was the outgrowth of the Judeo-Christian religious heritage, going all the way back to the time of Moses but emerging most aggressively in the Middle Ages. "By destroying pagan animism," White wrote, which had taught humans a respect for the power and spirit dwelling in the natural world, "Christianity made it possible to exploit nature in a mood of indifference to the feelings of natural objects." The Western religious tradition saw humans as the only species of moral significance on the earth and thereby sanctioned the uninhibited use, the misuse, even the wholesale extermination of the rest of the living world for the sake of satisfying human needs. Modern science and technology inherited from that religious tradition an attitude of indifference toward the intrinsic value of other forms of life, an attitude of militant anthropocentrism. To focus all the blame on contemporary technology for the crisis was to miss that profound moral conditioning that determined how technology was developed and used. The modern crisis, in other words, could not be explained as a mere deficiency in managerial skill among the technicians. More and better job training for them would not be enough, nor more and better tools. Rather, all people needed to think in less anthropocentric terms about our place in nature. We had to confront the powerful moral influence of Christianity and find an alternative relationship with the earth if human ecology was to escape its mounting crisis.

Historians like Lynn White never make things easier for others, for they tend to give big, abstract answers to questions that most people hope are concrete, uncomplicated, and quickly solvable. Couldn't we just recycle newspapers, we want to ask. Couldn't someone just give us a list of "fifty simple things we can do to save the earth"? No, White would have answered, we've got to do much more than that—do nothing less than reinvent our religion. We've got to think about the burdens of history, the deep, complex trap that traditional culture has left us in; we've got to question the ways we have learned to react to the world around us. It's a tough project.

As a fellow historian, I share White's ambition to dig deeply into the past to illuminate the present. But it seems to me that we don't have to look so far back as the Book of Genesis, nor do we have to indict the entire Christian heritage for our situation. We have a much shorter and distinctly *modern* cultural history to understand and fix. . . .

I believe the most important roots of the modern environmental crisis lie not in any particular technology of production or health care—the advent of medical inoculations, for example, or better plows and crops, or the steam engine, or the coal industry, all of which were outcomes more than causes—but rather in modern culture itself, in its world-view that has swept aside much of the older religious outlook. Let us call this modern culture by a simple name but think about it as a complex phenomenon: the world-view of *materialism*. It has two parts, economic and scientific, so intertwined and interdependent that even now historians have not fully probed their intellectual linkage. Together, the two parts forced a powerful cultural turn as important as what Karl Jaspers has called the "Axial Period" of human history, which occurred in the sixth and fifth centuries B.C., when so many of the world's great religious and philosophical systems took form—Confucianism, Buddhism, the pre-Socratics in Greece, the Old Testament prophets.[2] I see this new world-view—"post-Axial" we might call it—taking over western Europe in the seventeenth and especially the eighteenth century A.D., after a long spawning period, and manifesting itself in many so-called revolutions, including the Scientific, the Industrial, the Capitalist, all of which were only surface manifestations of a more fundamental change of thought.

For the biophysical world the more immediately significant impact came from the materialism that was economic. I mean the view that improving one's physical condition—i.e., achieving more comfort, more bodily pleasure, and especially a higher level of affluence—is the greatest good in life, greater than securing the salvation of one's soul, greater than learning reverence for nature or God. It encompasses the view that any individual's or people's success is best judged in terms of the number of their worldly possessions and their economic productivity. In current parlance, I mean worshiping the god of GNP. All through earlier history there were individuals who lived by a materialistic standard, but we cannot find any whole culture where materialism defined the dominant system of values until we arrive at the modern age, which is emphatically, unabashedly materialist in its ultimate goals and daily strategies.

This materialist revolution was also notable, I have hinted, for its *secularism.* That is, it was not motivated primarily by religious motives or visions; in fact it undertook to free people from a fear of the supernatural and tried to direct attention away from the after-life to this-life and to elevate the profane over the sacred. This secularized culture came to supersede not only Judeo-Christianity but almost all the other traditional religious and ethical systems of the world—not entirely but enough to make them secondary, marginal influences. A growing secularism put religious feelings on the defensive, even invaded the very core of religious expression, subverting and distorting it into many strange new forms, so that today we can find unembarrassed Hindu gurus buying fleets of Rolls-Royces or Protestant television evangelists selling glitzy condos in a religious theme park.

Materialist culture was also *progressive.* It repudiated the attitude toward time in traditional mentality, where preserving ancient, well-established cycles of nature and culture was considered one of the highest duties. Now, duty meant moving oneself and one's society ahead, escaping the patterns of the past, throwing off the dead weight of tradition. We call this the idea of progress and, though it has a moral or spiritual aspect, we think of progress mainly as an endless economic or technological improvement on the present. Take the materialist core out of progressivism and it loses most of its appeal, its power over the imagination, its driving force. . . .

And then, appearing as a third characteristic, this new world-view of materialism armed itself with an all-sufficing, elegant, self-reliant mode of thought called *rationalism,* which was supposed to take the place of authority or spiritual revelation. Rationalism taught a new confidence in the ability of human reason to discover all the laws of nature and turn them to account. It emphasized the inherent capability of average men and women to discover the right principles of action, or at least to discover their own enlightened self-interest and act accordingly. This new rationalism urged people to overcome their self-doubt and humility and strike out on their own. Pride, lust, gluttony, envy, selfishness, and greed all vanished as sins or were redeemed from their sordid past by the alchemy of reason and thus—extraordinary transvaluation!—became acceptable as the very engines of progress.

I have said that there are two dimensions to materialism as a world-view: economic and scientific. The latter is absolutely essential to the former, and may even be the prerequisite for its existence. This other materialism is the philosophy that nature is nothing but physical matter organized under and obeying physical laws, matter rationally ordered but devoid of any spirit, soul, or in-dwelling, directing purpose. On this view of nature converge many of our modern university departments of learning along with our extra-academic institutions of research and development, governmental bureaucracies, and multinational corporations, all of which tend to approach nature as nothing more than dead matter.

Historians of ideas point to the French philosopher René Descartes as the chief prophet of scientific (or mechanistic) materialism, for it was he who laid the foundations for the modern mechanistic perspective in both physics and biology. One of Descartes's main assumptions was that animals and plants are mere machines, constructed from material particles and somehow arranged to conform with the mathematical laws of motion: mere clocklike apparatuses, capable of complex behavior but lacking souls. In a way that no truly traditional Christian, believing in the sanctity of God's creation, could share, Descartes looked on nature simply as raw material to be exploited by the human brain. The aim of modern science, he argued, is to "know the power and action of fire, water, air, the stars, the heavens and all the other bodies in our environment, as distinctly as we know the various crafts of our artisans; and we could use this knowledge—as the artisans use theirs—for all the purposes for which it is appropriate, and thus make ourselves, as it were, the lords and masters of nature."[3] It is a dream that resonates down through the centuries, promising an intellectual conquest of mind over matter that knows no bounds.

Descartes did his most influential work in the second quarter of the seventeenth century, but even before him, in the first quarter of that century, another philosopher of science, the Englishman Francis Bacon, made even more explicit the link between the two halves of modern materialism. Scientific materialism, Bacon promised, would provide the means for improving the human economic estate—harnessing ideas to practical ends, thereby making us all rich beyond counting. Through active science, he promised, we could do more than sit passively in a seat of honor over the rest of creation, as Genesis had allowed; we could become creators ourselves, turning the rest of creation into power and wealth, using our reason to enlarge "the bounds of Human Empire, to the effecting of all things possible." . . .

Thus this new world-view—a materialistic outlook that was secular, progressive, and rational—stole onto the European scene, fought against the declining power of the Church and feudal order, and eventually won over the leading minds of the era. This world-view was carried along in the minds of Europeans invading the New World, conquering and exploiting its riches. They had the vast treasure room of Africa in their sights too, and soon would open up India for the new empire of commerce and reason, founding trading posts at Bombay in 1661 and Calcutta in 1691. Everywhere they came upon civilizations of stunning beauty, but always it was a beauty embedded in outworn religious and philosophical systems, which put the highest human value on the immaterial and spiritual. Those backward foreign peoples all seemed to be slogging along in

ignorance of the great material possibilities that lay around them—
the potential of their lands to produce inexhaustible wealth. . . .

❖   ❖   ❖

All these ideas have been so often and so well studied that it seems
a little trite to insist on them here, yet even today the profound
environmental consequences in that shift to materialism, or even
the very fact of the shift, are not understood widely enough, nor do
many people, in reading about the disappearance of tropical rain
forests or the disposal of toxic wastes, stop to realize that these current problems have their roots in a cultural turn that began centuries ago and often in a land far away.

Such cultural shifts do not, of course, come full-blown from the
mind of a single man or woman, but rather indicate deep, nearly
simultaneous shifts in the minds of thousands, even millions, of
people—whole civilizations suddenly taking off in unison like a
flock of geese migrating to the north country, wheeling and dip-
ping in close formation as though wired together. A single great
mind, however, can reveal the general direction in which the flock
is flying and draw the map that others are following by instinct.
The individual who more than any other served that function for
the rising materialist world-view was an English-speaking philosopher and scientific economist, Adam Smith. I nominate him as the
representative modern man, the most complete embodiment of that
cultural shift; and recommend that it is he, not Moses, whom we
must understand if we are to get down to the really important roots
of the modern environmental crisis. Robert Heilbroner has spoken
almost irreverently of "the wonderful world of Adam Smith," but
in his time Smith was indeed a wonderful visionary, as he remains
for many today who are just now discovering his logic and perspective. So how did Adam Smith look on the world around him?
Where did nature fit into his thinking? What were the long-term

implications that his ideas had for the natural order of Planet Earth?

A large homely fellow with a bad twitch and an absent-minded air, Smith was a most unlikely looking leader for any intellectual revolution. He was born in 1723 in the seaside town of Kirkcaldy, directly across the Firth of Forth from Edinburgh, where he grew up among fishermen and smugglers with the smell of salt air in his nostrils. After university studies in Oxford, a teaching post in Glasgow, and travels as a gentleman's tutor in France, he returned home in his mid-forties to Kirkcaldy and, living unmarried with his mother, devoted himself to writing his great book, *The Wealth of Nations*, published in that revolutionary year 1776. Although he is described as one who liked to take long solitary walks along the seashore, he never actually expressed any love of the sea or admiration of its beauty, never seems to have watched with any interest a gull hovering in the air, a crab scurrying over the rocks, the tide moving in and out. And though he lived in a Scotland that had severe ecological problems caused by overgrazing, deforestation, and soil depletion, he never considered how the Scots might change their land-use practices and become better stewards of their patrimony. And though many of his contemporaries were enthusiastic naturalists—it was a fabulous age of natural history, including the remarkable Gilbert White of Selborne, Carolus Linnaeus of Sweden, compt Georges-Louis Leclerc du Buffon of France—Smith seems to have lived his entire life utterly oblivious of the nature around him. He set out to revolutionize the study of human economics in total disregard of the economy of nature.

What Smith knew and thought about was the expanding life of commerce and industry, the rising class of businessmen, the mind of the entrepreneur, the factory system of production, most of which was found far from Kirkcaldy. Instead of moving to the very centers of commerce where he could make great bundles of money for himself, he chose to stand aside and observe, to see how it was done by others, and to help his nation, Great Britain, figure out how wealth had been and might be gained. Ironically, he was a humanitarian, a disinterested materialist who celebrated the amoral pursuit of self-interest.

The secret to increasing the wealth of nations turned out to be rather simple, though it took Smith an enormous body of text to reveal it. A nation that seeks wealth, he concluded, must establish a "system of natural liberty" in which "every man, as long as he does not violate the laws of justice, is left perfectly free to pursue his own interest in his own way, and to bring both his industry and capital into competition with those of any other man, or order of men."[4] Note that Smith called this system "natural," for he believed it was in harmony with the laws of human nature. It is natural, he believed, for humans to want, above all else, to increase their material comforts, to add to their sum of riches by "truck[ing], barter[ing], and exchang[ing] one thing for another." If that truly is the way all people naturally behave, then a society or culture would itself be most natural when it allowed, or even encouraged, people to enjoy as much freedom as possible in pursuit of their acquisitive natures. Smith did add what is overlooked by many of his later disciples: that a society may also rightfully restrain those human natures in the interest of social justice, but such restraint should not interfere too much with private freedom. To attempt to legislate a general benevolence, he believed, would be to try to overturn the laws of nature.

So little did Adam Smith consider what most people, then and now, mean by nature—the flora and fauna, the soil and water—that we cannot really speak at length about his philosophy of the subject. This much can be said: he did not conceive that the non-human realm lays any obligations on humans. What Christians called the Creation, what their religion required them to respect as the handiwork of God, had become for the economist quite value-less in and of itself. Value, in his view, is a quality that humans create through their labor out of the raw materials afforded by nature. A thing has value only when and if it serves some direct human use ("value in use") or can be exchanged for something else that has value ("value in exchange"). One of Smith's most influential predecessors, John Locke, declared that "the intrinsic natural worth of anything consists in its fitness to supply the necessities or serve the conveniences of human life."[5] He meant that nothing in the unimproved natural world has any intrinsic worth—a worth in and of itself—but only an instrumental worth, measured by whatever human uses it can serve. Likewise for Smith, nature is only instrumental and has worth or value only to the extent it has been "improved" by human labor.

The wealth indicated in *The Wealth of Nations* does not include any of the material benefits that humans derive from unimproved land: the air and water that sustain life, the process of photosynthesis in plants, the intricate food chains that we draw on for sustenance, the microorganisms that decompose rotting carcasses and return them to the soil. In a passage from the chapter "The Employment of Capitals," Smith does refer passingly to a "nature" that "labours along with man" in agriculture, adding fertility to the soil just as servants and domestic animals add their labor to improving the master's property. "Though her labour costs no expence," he writes, "its produce has its value, as well as that of the most expensive workman." In another passage dealing with Columbus's discovery of the New World, he indicates that "the real riches of every country" are "the animal and vegetable products of the soil," but then adds that Columbus found little wealth on his voyages but cotton and gold, dismissing even the Indian corn, yams, potatoes, and bananas the Italian brought back to Europe as unimportant economically. . . . Unimproved nature was for Smith a "vulgar" show, unworthy of a great man's interest. . . .

❧　❧　❧

The worldly philosophy of Adam Smith has become the dominant one in all the industrial nations, from Great Britain and Germany to the United States and Japan (the bulk of Smith's library now resides in Tokyo). So also in the nations that have followed, however faithfully or not, the teachings of Karl Marx, including the late Soviet Union. Marx may have been a sharp critic of the Smithian model of promoting economic growth through market freedom, but like Smith he was emphatically a materialist: secular, progressive, and rationalistic to the core, a fierce critic of all the traditional religions, all forms of pagan animism or Christian superstition, all reverence toward the earth. Marx and the Marxists were radicals for social justice, but they had the goal of material abundance firmly in mind too and were devoted to the modern world-view.

Now every economy in history, from that of the Bushman of Australia to that of global capitalism, has tried to extract resources from nature and turn them to human advantage. But no economy finds those resources in a void; they all must come out of a larger order or system. We can call that larger order "the economy of nature," following the lead of Smith's neglected contemporaries, the eighteenth-century naturalists. In this light every economy that humans have devised must appear as only a dependent economy, deriving from that greater one. We have not invented nature's economy; we have inherited it through eons of evolution. We learn to take things out of it for our own use and circulate them for a while within our little economy, turning forests into houses and books before yielding them to rot and mildew. The human economy requires for its long-term success that its architects acknowledge their dependence on the greater economy of nature, preserving its health and respecting its benefits. By this standard every modern economy, whether built on the principles of Adam Smith or Karl Marx, is an unmitigated disaster.

Once we acknowledge that the economy of nature is real and indispensable, then this entire modern way of thinking appears in a withering light as overweening pride in inadequate intelligence and skill. Living by overconfident materialism, people come to believe that they can create all the fertility they need by adding to the soil a bag of chemicals, that they can create any amount of wealth out of the most impoverished landscape, that they can even create life itself in a glass tube. To be sure, human artifice has improved our power over the elements, suggesting that nature's economy does not set rigid or fixed limits to our existence. But now we are learning that we cannot use that power as safely as we thought. We cannot anticipate all the consequences of our ingenuity, and greed, no matter how rationalized, remains the root of evil and self-destruction.

If my argument is right and the environmental crisis is really the long-preparing consequence of this modern world-view of materialism, economic and scientific, then it makes no sense to blame any of the traditional religions of the world. Religion, on the whole, acted to check that materialism, to question human arrogance, and to hold in fearful suspicion the dangerous powers of greed. Religion, including Christianity, stood firmly against a reductive, mechanistic view of the world. It pointed to a subordinate and restrained role for humans in the cosmos. And, most importantly for the sake of the biosphere, it taught people that there are higher purposes in life than consumption.

The ecological crisis we have begun to experience in recent years is fast becoming *the* crisis of modern culture, calling into question not only the ethos of the marketplace or industrialism but also the central story that we have been telling ourselves over the past two or three centuries: the story of man's triumph by reason over the rest of nature. But having presented that argument, I cannot now recommend that we slip backwards in time and solve the crisis by reading the Bible or Koran again. It is not possible, or even desirable, to try to go back to a pre-modern religious world-view. We cannot so simply undo what we have become. For this reason I must once more disagree with Lynn White, who proposed that the world convert to the religious teachings of St. Francis of Assisi, the famous thirteenth-century Italian monk who embraced the plants and animals as his equals and beloved kinfolk. The idea of making

Franciscans of everyone in the world would be an ethnocentric and anachronistic solution to the modern dilemma.

So what can we do? What is the solution to the environmental crisis brought on by modernity and its materialism? The only deep solution open to us is to begin transcending our fundamental world-view—creating a post-materialist view of ourselves and the natural world, a view that summons back some of the lost wisdom of the past but does not depend on a return to old discarded creeds. I mean a view that acknowledges the superiority of science over superstition but also acknowledges that all scientific description is only an imperfect representation of the cosmos, an acknowledgment that is the foundation of respect. I mean the view that all consumption beyond a level of modest sufficiency is pathological in both a personal and an ecological sense; like any kind of gluttony it deserves pity, not approval. I mean the view that greed is always a vice, not a virtue, that unlimited economic growth or "development" has become a fanatical drive against the earth. Whether such a viewpoint might first appear in the most advanced industrial societies, where so many people have begun to have doubts about the world they have made, or in the least advanced, where most people are still converting to the modern notions, though with many doubts of their own, I cannot say; only that such a post-materialist culture must appear somewhere in embryonic form and spread eventually, as the doctrines of Adam Smith have done, to the farthest corners of the earth.

PART III

SHARING LIFE

Chapter 12

Great Possessions

David Kline

Although the agrarian persuasion characterizes and motivates people in all walks of life, its essential guiding image remains the farm family, living and working on a small, diverse, ecologically sound homestead. In such a setting, the connection to the land is plain, immediate, and full.

In his regular contributions to farm magazines, Amish farmer David Kline describes the joys that today can accompany a thoughtful, healthful life on the land. With his family and neighbors, Kline farms 120 acres of land in northern Ohio. This chapter begins with one of Kline's nature sketches, "A Spring Walk," drawn from his engaging essay collection Great Possessions: An Amish Farmer's Journal. *Like many agrarians before him, Kline is an attentive naturalist and delights in the comings and goings of his wild neighbors. We see in this sketch the joy the Klines take in sharing the land not just with familiar human neighbors but also with birds, mammals, mushrooms, and flowering trees. The selection continues with the introduction to Kline's book, a more sober piece in which he explains his family's practices and principles and expresses his perplexity about mainstream American culture. In it, one sees how linked his family is to the surrounding human community as well as to the land.*

From slightly fewer than 4,000 members in 1900, the Old Order Amish population in the United States has risen to more than 170,000, with settlements in twenty-two states. To various degrees, its living practices are also embraced by members of Mennonite churches and other pietistic religious organizations.

Last May our youngest daughter, Emily, then eight, and I carefully planned an outing. Her older sisters dropped us off several miles southwest of our farm on their way to Berlin. We then intended to walk the seven or eight miles to Millersburg, catch the evening bus, and ride home. This route took us through what I consider some of the most beautiful woodland country in this part of the county. Besides, we crossed some of the farms of friends, giving me a chance for a short visit. We traveled light. Only one pair of light-weight binoculars, a field guide to wildflowers, a Sears sale catalog to serve as a makeshift plant press, a quart of water, some snacks, and a breadbag for gathering edibles. Since the day started out foggy we waited until the warming sun dissipated the fog. By then the temperature reached the 60s. A perfect day for walking. We had no particular subjects in mind. As Emily said, we just went see-ing things.

As we left the road and walked downhill toward the woods, our attention was drawn to things yellow. The lush green pasture we were walking through was splattered profusely with blooming dandelions. From a wild cherry tree several goldfinches took flight, the males handsome in their yellow-and-black coats. Then a bril-liant yellow warbler sang his cheerful song from the top of a thorn apple, while from a nearby tangle of blackberries and multiflora roses came a spritely "witchity-witchity-witchity," the song of the common yellowthroat. This black-masked, yellow-breasted song-ster is a warbler, yet its habits are almost wrenlike, as it flits low through the underbrush.

Nearer to the woods a loud unmusical song burst forth from a patch of briars. It sounded like a brown thrasher or mockingbird, because the singer even mimicked a crow. But as the bird flew from the thicket and perched on a greenbriar cane, we saw, to our sur-prise, that the mimic was a yellow-breasted chat.

The chat, too, is a member of the warbler family, even though it is bigger than the other members of the clan, and its ways aren't very warblerish. In beauty, though, it belongs to this colorful family so beloved by birders.

Our walk was off to a good start.

After getting a good look at the chat we went on our way. We hadn't gone far when we surprised a woodchuck basking in the sun. We laughed as the startled rodent scrambled through the branches of a fallen tree to dive into the safety of its burrow, which was more than likely connected to the cavernous tunnels of an abandoned coal mine nearby.

While crossing the fence to the woods, Emily spotted two deer. They must have seen us coming, yet the shy animals didn't seem to be greatly alarmed. After a while the deer seemed to have satisfied their curiosity, and with a waver of their white tails they bid us farewell.

Beneath the canopy of the mature maples and oaks we listened to the songs of the eastern wood pewee, scarlet tanager, and red-eyed vireo, and in the distance the clear call of the rufous-sided towhee—"chewink! chewink!" The male towhee is a bird of contrasting colors. Sporting an ebony-black head, throat, and back, along with rufous sides almost matching the color of a robin's breast, and a white belly, he is a dashing bird. Towhees nest and feed on the ground. They often are seen scratching in the leaf litter for food, hence their local name, ground robin.

Coming to a long-unused field, we stood in awe, dazzled by the display of blooming dogwoods. We picked one flower, still wet with dew, which measured over four inches across. The dogwoods weren't the only spectacular thing about the field: there were also bluets. These tiny four-petaled flowers were growing and blooming in such abundance that, at a distance, the field appeared frost-covered. Reversing the binoculars and using them as a magnifying glass, we took a close look at the blue flower with the golden eye. Also called quaker ladies and innocence by some, they are very pretty. As a farmer, I must admit that the neglected field with its clumps of poverty grass bothered me. But what a grand way for a field to be returning to woodland.

Reentering the woods, we followed a ravine down to the bottom and then walked alongside the winding creek. We found a few morel mushrooms and a new flower here in the rich bottomland. The flower was in the orchid family, and the field guide showed it to be a showy orchis. Though the lavender and white flowers weren't as large as the related lady's slipper, we were nevertheless excited at finding one member of this beautiful family of flowers.

We continued along the creek, checking around sycamores for mushrooms and watching the trees for birds, until we came to a place I remembered from some years before to be home to large numbers of wild leeks. Using a stick and our hands we dug out several dozen of the pecan-sized bulbs. The leek is in the onion family, and this early in the spring I consider them a delicacy. Dropping them in the bag with the mushrooms I could almost taste the upcoming meal—fried mushrooms and sliced leeks on a warm piece of freshly buttered bread. After we gathered the leeks, we sat down and ate our snack.

Coming to Martin's Creek, we took off our shoes and waded across rather than going out of our way to cross a bridge. The cool water was refreshing, and before going on we rested until our feet were dry. Walking along an abandoned township road south to the next valley, we found two more flowers we had never seen—spring larkspur and dame's rocket.

After visiting with a friend, we crossed his farm and climbed the next ridge. Reaching the top, we saw an American redstart, which surprisingly was one of the few migrating warblers of the day. The perfect weather that had allowed us to get the oats and corn in ahead of time probably also contributed to the dearth of migrating warblers. With no storms to force the birds down, they overflew this area on their northward journey. At least I hope that is the reason. I fear, though, that the cutting down of the tropical rain forests (the winter home for many warblers) to create ranches that will provide cheap beef for fast-food restaurants in the United States may also be partly responsible for the dearth.

Dropping down the other side of the ridge, we came to a plateau of about an acre, shaded by majestic beeches and maples, that was solid with blooming white trilliums and lavender wild geraniums. Off to the left a Kentucky warbler began his rollicking song. The scene was simply idyllic. At times like this I want to remove my hat to the beauty of the natural world and its Creator. While we sat on a log and admired, I checked my watch. Four thirty. Exactly the time the bus was to leave Millersburg, and we were still several miles away. Emily looked at me, shrugged her shoulders, and said, "Oh, well."

Crossing the deep ravine, we found wild ginger on what was now one of our good friend John Y's farms. Along the edge of one of the fields a loud bird song interrupted our trek. I had heard the song before but couldn't recall its owner. The bird cooperated by revealing himself, and we identified him as a white-eyed vireo. Of all the vireos the white-eyed is the least accomplished songster. He doesn't sound as if he were even related to the warbling vireo.

We were now on "solid ground" again, a shaded lane. Here we heard and saw a rose-breasted grosbeak and an orchard oriole. And though we didn't keep a list of all the birds we saw—it probably wouldn't qualify as a big day—we were satisfied.

At the end of the long lane we decided to walk the rest of the way to town on a blacktop road. We soon realized we were back in civilization as NO TRESPASSING signs began appearing on trees and poles. We abided by their wishes. The last mile or two on macadam road tired us more than all the miles through woods and fields.

After a quick sandwich and milkshake, we called a friend and were given a ride home, tired (especially Dad) and happy and accompanied by the aroma of wild leeks. . . .

❀ ❀ ❀

Not too long ago, the editor of a back-to-the-land magazine asked
me to write something on small-scale diversified traditional farming—on the advantages of such a way of life, he suggested, and also
the disadvantages. This bothered me all summer. Quite honestly, I
couldn't think of any disadvantages.

What are the lessons, if any, I wondered, to be learned from our
way of farming? Is it a way of farming that preserves the soil, the
water, the air, the wildlife, the families that work the land, and the
surrounding communities? In other words, are we proper caretak-
ers or stewards of God's Creation? Are we in harmony with God
and nature?

To write about Amish agriculture is to write about traditional
agriculture, an agriculture dating back to eighteenth-century
Europe, handed down from generation to generation and yet with
innovations and improvements constantly added along the way.
The Amish are not necessarily against modern technology. We
have simply chosen not to be controlled by it.

Amish farming is sometimes best looked at by someone outside
the community, for many of our practices are so traditional, having
been handed down from parents to children for so many generations, that the reasons are almost forgotten. For example, the rotation of our field crops here in eastern Ohio works so well it's seldom
questioned. This is a four- or five-year rotation, which means a
given field will be in corn every fourth or fifth year. (I should mention that the type of Amish farming I'm talking about is practiced
by our people in Ohio, northern Indiana, southern Michigan, and
possibly southern Ontario. Amish communities in other states may
differ somewhat in their ways of farming. Yet there are many sim-
ilarities, I'm sure.)

In our rotation, corn is followed by oats. In the fall after the oats
are harvested, the stubble is plowed, and wheat is sowed; the wheat
is then top-seeded the following March or April with legume seeds.
Seeding is done with a hand-cranked or horn-type seeder and usu-

ally on frozen ground, where the early-nesting horned larks nest. The dropping seeds cause enough disturbance to flush the incubating bird. The nest is then easily found.

After the wheat is cut and threshed in July the stubble is mowed and, almost miraculously, the wheatfield converts to a hayfield. The next spring and summer several cuttings of hay are made and then the hayfield is pastured in the fall. (In a five-year rotation the field remains in hay for two years.) During the winter the old sod is liberally covered with strawy manure. In late winter or early spring the sod is plowed and in May planted again to corn and the rotation or cycle begins again.

What chemicals must be bought and added to raise a decent crop of corn in a field like this? None, except for the fungicide with which the seed corn is usually treated before we buy it. With the legumes converting free nitrogen to the soil from the air plus ten to fifteen tons of manure per acre supplying thirty pounds of nitrogen per ton in addition to other plant foods, no extra fertilizer is needed. (I should add that the majority of Amish farmers apply one hundred to one hundred and fifty pounds of a low-analysis fertilizer—such as 5-20-20—per acre as a starter plant food. I know quite a few farmers, though, who don't use any purchased fertilizer for corn and raise an excellent crop.)

Likewise, no insecticides are needed in this field because with corn following hay there are no crop-damaging insects. We have never used a soil insecticide. Rarely, in wet weather, slugs will cause some problems in corn planted in plowed sod. But the first cultivation destroys their burrows, and by the time the slow creatures get geared up for their next attack, the corn has outgrown the stage where it can be damaged.

The cultivating, besides taking care of the slugs, also takes care of the weeds. Most Amish farmers are not what you would consider pure organic farmers. Many will use some herbicides in corn to help control problem weeds and grasses. But the Amish usage of herbicides is small in comparison to that in conventional continuous corn farming because the need is small. One year my neighbor's total herbicide cost was $11—not per acre, but for *all* his corn. He band-sprays a small amount on the row. Between the rows he gets the weeds with his cultivator.

Most of us aren't too concerned if there are some weeds and grasses in our corn. In fact, I want some there. Occasionally we get summer thundershowers that dump several inches of rain in half an hour or less, which is more than even the most absorbent soil can take. During storms like this we depend on a smattering of quack grass and on sod waterways to hold the topsoil.

According to researchers at Oberlin College, the strength of our topsoil can be attributed to the tilth of our soils. Their study shows that our traditionally horse-worked farms absorb almost seven times more water before becoming saturated than the conventional no-tilled farms.

Presently no-till farming with its dependence on vast amounts of chemicals is being touted by the experts as the way to guarantee green fields forever. What they fail to say is that these green fields will be strangely silent—gone will be the bobolink, the meadowlark, and the sweet song of the vesper sparrow in the twilight.

A young farmer friend related to me how thrilled he was last spring when two pairs of bobolinks took up residence in one of his sod fields, a field that was to be planted to corn using the no-till method. He hesitated somewhat to spray the field, but then he remembered the film the Chevron salesman had shown the previous winter indicating that no-till improves the habitat for wildlife. He sprayed the paraquat and soon after, the bobolinks disappeared. Since he didn't hear their cheerful flight songs from nearby fields he's quite positive the birds perished.

Another disadvantage with no-till is that the farmers' options are severely limited. Should there be too much or too little rain—conditions which often reduce the effectiveness of many herbicides—or an invasion of army worms or slugs, the farmer can't cultivate and must instead come back with more pesticides. I was recently told that during late spring and early summer every rain-drop in the eastern corn belt contains minute parts of *Lasso,* a pop-ular corn herbicide and suspected carcinogen. Can you love your neighbor and do this?

But, to champions of agribusiness, that is progress. And profits ... for them. One Soil Conservation Service (SCS) board member made the comment, "The Amish minds are too 'unscientific' to understand the intricacies of proper soil management, so they should learn to rely on outside experts for advice." At another farmer-advisor meeting when the discussion drifted around to no-till, the expert, educated in the atmosphere of a land-grant college where the jargon revolves around "input," "output," "acre-eaters," "work-is-drudgery," "cash flow," and "bottom line," made the remark, "No-till sure beats plowing."

Here it is then, the thorn in my side—I never did care for Faulkner's *Plowman's Folly*—I enjoy plowing. Just this past year the SCS technician told me, in all seriousness, that if I'd join the no-till crowd I'd be freed from plowing, and then my son or I could work in a factory. He insinuated that the extra income (increased cash flow) would in some way improve the quality of our lives.

I failed to get his point. Should we, instead of working the land traditionally, which requires the help of most family members, send our sons to work in factories to support Dad's farming habit? Should we be willing to relinquish a nonviolent way of farming that was developed in Europe and fine-tuned in America (by what Wendell Berry calls "generations of experience")? Should we give up the kind of farming that has been proven to preserve communi-ties and land and is ecologically and spiritually sound for a way that is culturally and environmentally harmful?

And there are the pleasures of plowing—plowing encompasses more than just turning the soil. Although I can't fully describe the experience, it is like being a part of a whole. In early spring, my son and I, each with a team as eager to be out as we are, turn the mellow soil, feeling its coolness and tilth. We take pleasure in the transient water pipits and pectoral sandpipers feeding on the freshly turned earth abounding with life. As we rest the teams, I listen to the joys and uncertainties of teenage years.

Maybe I'm blind, but no matter which angle I look from, I fail to see any drudgery in this work. And I am convinced that if one farms carefully, soil erosion need not be a problem.

Several springs ago—actually it was in late winter—following a week of unseasonably warm weather, Dennis Weaver, our neighbor to the south, couldn't resist the urge any longer and started plowing. I wasn't aware of it until, while walking to the barn, I suddenly caught the aroma of newly turned earth. I stood there, closed my eyes, and reveled in it: the promise of spring.

With no-till I would have the means to farm his fifty tillable acres, in addition to my own, and he could be "free" to work off the farm. I know I wouldn't be able to do the excellent farming he is doing now, and I would miss the rich fragrance of his fertile soil. But more than that, I would miss my neighbor.

There are lessons to be learned from small-scale diversified farming. By working and farming the way the Amish traditionally have done, we make our place more attractive to wildlife. Should we be removed from the land and our farm turned into a "wildlife area," I'm almost positive that the numbers and species of wildlife would dwindle.

Naturally, this doesn't mean that farming to the roadsides and cleaning and spraying fencerows, as even some Amish do, particularly in communities of high land prices, can be done without ill effect to wild things. Along with a diversity of crops and livestock and minimal use of pesticides, there should be some overgrown fencerows—which harbor a host of wild creatures from catbirds to cottontails—brushy woods' edges, sod waterways, trees around the farm buildings, an orchard, lots of flowers (both garden and wild), maybe a patch of prairie.

Gary Nabhan wrote in *The Desert Smells Like Rain* about two Sonora Desert oases, the first of which, in Arizona, began to die when the Park Service turned it into a bird sanctuary and, in an effort to preserve it for wildlife, removed the Indians who farmed and lived there. The other oasis, across the Mexican border, has long been tended by a village of Papago Indians and is thriving. An ornithologist found twice as many species of birds there as he found at the bird sanctuary in Arizona.

Last week our family made a survey of nesting birds around our farm buildings. This doesn't include the bobolinks, redwings, meadowlarks, and sparrows in the fields, nor does it include the vireos, tanagers, warblers, and thrushes in the woods or the rough-winged swallows and kingfishers along the creek. We came up with over eighteen hundred young of thirteen species fledged within two hundred feet of our house. This included a colony of 250 pairs of cliff swallows along the barn eaves. As Mr. Nabhan's Indian friend said, "That's because those birds, they come where the people are. When the people live and work in a place, and plant their seeds, and water their trees, the birds go live with them. They like those places, there's plenty to eat, and that's when we are friends to them."

But we farm the way we do because we believe in nurturing and supporting all our community—that includes people as well as land and wildlife. By farming and living independently of electricity the Amish are not contributing, at least not directly, I hope, to the destruction of hundreds of farms and communities in southeastern Ohio where the Ohio Power Company is strip-mining coal to supply its power plants on the Ohio River. Along with the destroyed farms, the mammoth power plants spew out sulfur dioxides that contribute to the acid rain killing forests in the Northeast and lakes in the Adirondacks.

The Amish have traditionally maintained a scale of farming that enabled each farm to be worked by a family. Few farms have more than eighty tillable acres, which is about the maximum a father and son can easily work. If more help becomes available the operation may be expanded to include more livestock or possibly specialty crops such as vegetables. Rarely are more acres added.

Wes Jackson has said, "The pleasantness or unpleasantness of farm work depends upon scale—upon the size of the field and the size of the crop." The Amish have maintained what I like to think is a proper scale, largely by staying with the horse. The horse has restricted unlimited expansion. Not only does working with horses limit farm size, but horses are ideally suited to family life. With horses you unhitch at noon to water and feed the teams and then the family eats what we still call dinner. While the teams rest there is usually time for a short nap. And because God didn't create the horse with headlights, we don't work nights.

We have seventy tillable acres, which is maybe ten acres more than the average farm in our community. We couldn't take care of more. With this size farm there is usually something to do, yet we're never overwhelmed by work. I confess, though, that this past July we came close to being overwhelmed. Rain delayed the second cutting of hay and then when the rains quit we had hay to harvest, wheat to thresh, and oats to cut, all at once. Under normal conditions, though, the work is spread out from spring to fall.

The field work begins in March with the plowing of sod. This is leisurely work, giving the horses plenty of time to become conditioned, and giving us my version of what the Quakers call quiet time: a time to listen to God and His Creation as we participate in the unfolding of spring. Wendell Berry writes in *Getting Along with Nature:* "A proper human sound . . . is one that allows other sounds to be heard"—and plowing sod is such a human sound. We hear the creaking of the harness and the popping of alfalfa roots, as well as the tinkling song of the horned lark and the lisping of the migrating water pipits. A wonderful time.

April is for plowing corn stalks and sowing oats, for spring beauties and lovely hepatica.

In May, we plant the corn, turn the cows and horses out to pasture, and revel in warblers and morel mushrooms.

With hay making in June come the strawberries, shortcakes, pies, and jams. The bird migration is over and summer settles in.

Life and work on the farm peaks in July with threshing, second-cutting hay, transparent apples, new honey, blackberries, and the first katydid.

August already hints of autumn. The whine of the silo fillers is heard throughout the land. We fill our ten-by-forty-foot silo with the help of four neighbors.

McIntosh apples and sowing wheat mean September.

October is corn harvest and cider-making and loving the colors and serenity only October can offer. As the month draws to a close so does the field work.

The year is a never-ending adventure. What many consider recreation we enjoy on our own farm. This year we saw four firsts on the farm: our first Kentucky warbler; our first luna and imperial moths; and after waiting for over thirty years, I saw my first giant swallowtail butterfly.

The aesthetic pleasures of diversified farming are obvious. From spring through fall the colors of the fields are constantly changing. I like to look at our farm as an artist would behold his or her painting—a variation of colors and designs, never a bare spot of canvas left exposed. The bare spots on our farm, such as cow paths, are covered in November with strawy horse manure to prevent erosion. I use the manure spreader, which works fine as a mulcher. The land is now ready for the rains and storms of winter.

Probably the greatest difference between Amish farming and agribusiness is the supportive community life we have. Let me give an example. When we cut our wheat in early summer (we cut about half of a thirteen-acre field in one day), the whole family, after the evening milking, went shocking. It was one of those clear, cool June evenings. Simply perfect. Tim, our eighteen-year-old son, and I each took a row while my wife, Elsie, and ten-year-old son, Michael, took another one. Two of our daughters, Kristine, sixteen, and Ann, twelve, took the fourth row. Eight-year-old Emily carried the water jug. Row by row we worked our way across the field, the girls talking and giggling while they worked and Michael explaining in excited detail some project he had under way in the shop. When we reached the top of the hill we stood together and watched the sun slip behind a brilliant magenta-colored cloud and then sink beneath the horizon. From far to the south came the mellow whistle of an upland sandpiper. Tim said, to no one in particular, "Shocking together with the family is fun." He spoke for all of us. Then we heard voices from the next hill and saw three neighbors shocking toward us from the far end of the field. One of the girls excitedly remarked, "Seven rows at a time. That is speed." Soon all the bundles were set up in shocks and everyone came along to the house for ice cream and visiting.

The assurance and comfort of having caring neighbors is one of the reasons we enjoy our way of farming so much. Eight years ago I had an accident that required surgery and a week in the hospital. My wife tells me the first words I said to her in the recovery room were, "Get me out of here; the wheat has to be cut." Of course, she couldn't, and I need not have worried because we had neighbors.

While Dad cut the wheat with the binder, the neighbors shocked it. When our team tired my brother brought his four-horse team, and by suppertime the twelve-acre field was cut and shocked.

This year the neighbor who had been in first to help us needed help himself. Since a bout of pneumonia in July he hadn't been able to do much. So last Thursday six teams and mowers cut his eleven acres of alfalfa hay. Then on Saturday afternoon, with four teams and wagons and two hay loaders and fifteen men and about as many boys, we put the hay in his barn in less than two hours. We spent almost as much time afterward, sitting in a circle beneath the maple tree with cool drinks and fresh cookies, listening as one of the neighbors told of his recent trip west. He and a friend visited draft-horse breeders in Illinois, Iowa, and eastern Nebraska, and what a story he had to tell: of nice horses and nice people, of the worst erosion he had ever seen from the Iowa hills following eight inches of rain, and how the Iowa farmers rained invective down on our president. "Ach," he said. "All they want is more government handouts."

I couldn't help thinking of my young friend who got married last September and then bought his dad's machinery and livestock and rented the farm. He and his wife really worked on that debt. Milking by hand, selling Grade B milk, tending a good group of sows, cultivating corn twice, some three times, using no herbicides, they are nearing the end of their first year of farming on their own, and most of their debts are paid off. He didn't tell me this, he's much too humble, but he did say to me while threshing, "You know, farming is good."

Chapter 13

Reclaiming the Commons

Brian Donahue

*Although the diverse, privately owned farm supplies the guiding image
for the agrarian persuasion, agrarians necessarily look beyond the single
farm or parcel to consider the needs of larger landscapes. What patterns
of landownership and land use are needed to sustain the landscape as a
whole, ecologically, economically, and aesthetically? What communal
structures and educational programs are most likely, in a given place
with a given history, to foster social bonds and stimulate responsible,
communal behavior?*

*Few writers have spoken more sensibly about this need for education
and action at the community level than Brian Donahue, whose work has
centered on the town of Weston, Massachusetts, a small outer suburb of
Boston. Together with his wife, Faith, and other community members,
Donahue organized and directed Land's Sake, a successful market-
garden operation that expanded gradually to include timber management, orchards and maple-syrup making, a sheep herd, and various land-
centered education programs. Land's Sake insisted, and still insists, that
its programs make practical sense ecologically, economically, and aesthetically. It also insists that these programs engage local residents, entic-
ing them out of their homes and cars and encouraging them not just to
see the land anew but also to grasp the possibilities for rebirth of local,
sustainable, land-based economies.*

*On the basis of his experiences, Donahue concludes that towns such as
Weston can flourish best if many lands are owned in common and managed by the local community. Donahue supports private ownership;
what he criticizes is its excess, the fragmentation of people, lands, and*

perceptions that occurs when a landscape is nothing more than a collec-
tion of discrete parcels, separately owned and separately managed. In this
selection, drawn from his book Reclaiming the Commons: Community Farms and Forests in a New England Town, *Donahue describes*
the needs of Weston and similar New England towns threatened by
sprawling suburbs. He ends by encouraging people everywhere to reclaim
their common lands and reconnect with their shared agrarian past.

The suburbs are coming. In many places the suburbs are here, and
in many more they are swarming just over the horizon, their vanguard leaping ahead along the ridgetops to command the best
views. I was born in suburbia, and I now fully expect to be buried
in it. I am not happy about this, but every generation has to meet its
fate. Americans are going to continue moving to rural places and
small towns. Unless we adopt better ideas about land ownership
and care, Americans are going to continue ruining these places and
bankrupting the future. Suburbanization is an impulse that
destroys the very rural character it seeks. Therefore, those of us
who think we have some better ideas about how rural land might
be reinhabited must consider how we can harness the suburban
impulse and turn it toward a better end.

Suburbia, which epitomizes and requires the profligate consumption of nature driven by industrial capitalism, is probably not
a sustainable form of settlement. I certainly hope not. . . . I have
summarized the reasons I believe it must be stopped, or rather
transformed: the necessity to reduce our rate of resource extraction
and excretion to what the biosphere can tolerate, the necessity to
grow food and wood in ways that protect our water and soil, the
necessity to safeguard the biological diversity, ecological health, and
beauty of the world that enfolds us. Understood in these terms, a
thriving rural world is necessary to our survival, but we are cur-
rently treating what is left of that world as a fleeting luxury to be
consumed by those who can race momentarily ahead of the rest.
This obviously can't go on forever. Unfortunately, the suburban

drive no doubt will be sustained for several decades at least—long enough to do much more damage. In one way or another, suburbanization will occupy and change many more rural places. We need to see to it that rural resettlement takes a new form that permits and even promotes better ways of living.

Champions of small farming like Marty Strange, Gene Logsdon, Wes Jackson, and Wendell Berry have long warned that we are losing our family farm culture and rural communities. For generations now, the bell has been tolling for agrarian America. Where the agrarian tradition still lives it is worth fighting to defend it. Deep in the American heartland are regions that continue to lose population, that desperately need the return of people to take better care of the land and make depleted communities whole again. However, it is now clear that the problem confronting many other rural places is not simply how to prevent decline of one kind, but how to survive growth of another. Carried along by cheap transportation and ubiquitous telecommunications, the affluent can now live wherever they choose. Having laid waste to the countryside within commuting range of the beltways, they covet second homes and vacation chalets in more remote places of particular charm. No hilltop in America is safe from them, as they seek refuge from themselves. The rural diaspora we agrarians have long dreamed of is taking place before our eyes, in the worst way we could ever have imagined. We have lived to witness rampant decentralization of consumption, paradoxically driven by continued centralization in the control of productive land and the extraction of natural resources.

This residential flood drowns local places, rather than nurturing them. Conflicts inevitably arise between those few who were born in rural areas and make their living by working the land with efficient industrial machinery and the nouveaux rustiques who understandably would rather not live next door to such enterprise, even though it ultimately feeds and houses them. Ironically, many of these rural newcomers consider themselves environmentalists, but their ideas about what constitutes a healthy relation with the land tend to be confused at best. Their often passionate desire to leave land near their homes in an imagined pristine state of nature is in stark contradiction to their voracious consumption of natural resources. This is duplicity in place of atonement.

Our society rewards naked exploitation of nature and indulges naive, romantic love of nature, but the middle ground of caring for nature while using it productively has been declared uninhabitable. Our economy drives us to use nature as cheaply and unattractively as possible, and to compensate, we pay extravagantly to enshrine natural beauty in a few selected unproductive parts of the landscape, ranging from showpiece yards to wild refuges. The wild areas cost us nothing but a little lost production; landscaping suburbia consumes enormous expenditures. There is an odd dualism, or even trebleism, at work in our attitude toward land here. We demand that land be either cheaply used, expensively manicured, or utterly untrammeled in order to fulfill separate functions. Any suggestion that we might farm and log in restrained, beautiful ways in the first place is dismissed as inefficient and nostalgic at best and a sentence to misery and starvation at worst. This is economically absurd, but that is our economy.

We need a new attitude toward the land that sustains us, what Aldo Leopold called a "land ethic" half a century ago. But how do we get it? Can we engender a new, broad-based agrarianism that embraces both rural natives and suburban newcomers, that both protects and uses the land, under such polarized circumstances? I believe that we can, and that such a *common agrarianism* is urgently required. It will be by nature a bit less efficient at extracting resources from the land but will provide other benefits that ought to have economic standing. The keys to it are, first, reclaiming our common interest in the land; and, second, creating local economies that fit the land well and that actively engage people who move into and grow up in rural communities. We need to develop a common culture of caring for land, and we need more common land under our care.

Here are two plied but distinguishable strands: the way we own land and the way we use land. I want to address first how to better balance private and common interests in owning land, and then how to build strong local economies and culture. This is a call to the commons. But in calling for more common land I am *not* advocating a sweeping collectivization of agricultural production. Commons systems are never founded on exclusively communal ownership of property, and state systems that have approached this extreme have been clumsy, brutal, and ugly. I am not proposing to turn the suburbs into agricultural communes, but I am tired of watching private greed trample common good at every opportunity. Neither am I proposing an atavistic return to medieval common field farming and all that went with that, from the oxcart to the Black Death. Living in such a world would make me sick—in fact, I would already be dead several times over. I am equally queasy, however, about the fate of soil, water, food, and forests under the free market colossus that has been advancing for five hundred years and that now confidently bestrides the globe. It is clear to me that exclusively private ownership of land and extraction of commodities in a market economy is a better-paved road to a bigger ruin. We need a mix of private and common interest in land that is appropriate to the world today, one that balances personal freedom and community responsibility, economic efficiency and ecological restraint. In my view, this means more commons; but it also means a strengthened role for small private owners and entrepreneurs, as opposed to large corporations—which call themselves private even though they are plainly oligarchic collectives. Corporations that are devoted to maximizing profit have no rightful business owning land, which needs to be cared for in ways that necessarily reduce profit. The call to the commons must also be a call to the private rights of commoners.

Where does the common interest in land lie? That depends on the place. The balance between private and common interests will obviously vary from one kind of land to another within a community, and from one community to the next. Some kinds of land are best held privately, some commonly, and some with a mixture of private and common rights. We can think of the common interest as ranging from regulations that limit the way land may be used, through easements that convey restrictions permanently to the community for safekeeping, to outright common ownership. In general, the more inhabited and intensively cultivated the land, the more it belongs in private hands, and the more uninhabited and lightly managed, the more it belongs in commons. That is, as we move from the household toward the wild, we should also move from the most private toward the most common.

Each community needs to determine an appropriate balance among these elements—there is no perfect formula. But we can do far better than we have been doing. Weston now has coming on one-quarter of its land in commons and most of the rest in house-lots, which is a great achievement given the town's proximity to Boston and the era during which it suburbanized. Towns such as ours that have endured half a century of intense, often white-hot development pressure are the crucibles in which new land values have been forged. That land may have common as well as private value is still an awkward young notion in our time, but at least it has been reborn. Other rural places must learn to embrace and nurture this value earlier in the struggle, to protect more land. What balance of ownership might be achieved in a town that is just beginning to suburbanize? For simplicity's sake, let me examine three categories that stand for the full spectrum of land uses: residential land, farmland, and forest.

First, residences. I have no quarrel with the American ideal of private home ownership. It is entirely appropriate that a community contain a substantial number of single-family dwellings. (It is also appropriate that a community prevent the rising value of land from pricing out the less affluent. A community needs to make home ownership affordable for most of its citizens and also to have other kinds of decent housing available. Weston has failed this responsibility.) I *do* have a quarrel with the extravagant five- and ten-acre lots that often now surround houses being built on the suburban fringe and with the far-flung dispersal that results. A more corrosive pattern of settlement could hardly be imagined. Such detached housing, as it is aptly termed, ends by expressing extreme individualism, gluttony, and ostentation. Trophy houses, we call them here. For a family to enjoy private space for a yard and gardens is fine. But as houselots grow in size and multiply across the landscape, farmland and forestland are fragmented and the commonly enjoyed rural character of the place is rapidly consumed. The common interest should be to keep residential (and commercial) development confined to the least possible space that will accommodate a healthy rural community of a few thousand people—let's say not more than one-quarter of the land area and preferably far less.

Reclaiming the commons requires a new vision of the benefits and responsibilities of private home ownership within a community, a vision that extends beyond the garden fence. It will require a dramatic mental shift for upwardly mobile Americans to be satisfied with close quarters in place of the spacious suburban lot. People do not give up their dream of detachment easily. Those moving into rural communities must come to believe that it is in their best interest not to occupy as large a piece of land as they can, but rather to occupy as small a parcel as they can in order to enjoy more neighboring farmland and forest instead. Each new building lot should be counterbalanced by a deposit of land into the commons. Rural newcomers must accept their responsibility to protect the pastoral character that attracted them in the first place and to actively support a strong local land-based economy. That is, with a small parcel to call their own, nonfarming residents need to recognize that living in a community with attractive working farms and access to common forests confers private benefits that outweigh engrossing a large, expensive miniestate of their own; and that they can't have both. For better or worse, the era of the elite—wealthy enough to afford large estates in rural towns close to cities—ended half a century ago. Today, a much larger, more mobile affluent class subdivides and destroys. But why divide a one-hundred-acre farm into five-acre farmlets when the same twenty houses and their gardens could be confined to one corner of the property, leaving the farm largely intact and functioning and in some measure open for all to enjoy? I will have more to say about nurturing such nascent agrarian aspirations among an increasingly residential rural population in a moment.

Farmland lies in the middle of the spectrum of private and common interests in land. I am all for private family farms. I am convinced that in time the economic advantage will return to those who work smaller pieces of land in more intensive, diversified, sustainable ways than the average commercial farm of today. I hold this view because I believe the cost of industrial inputs will rise, tolerance for environmental damage will run short, and demand for fresh, wholesome, locally grown food will continue to grow. Working the land intensively will require native farmers who know their place well and care for it with devotion, a breed we have seen far too little of in our history of occupation to date. Experience has taught me how long it takes to get to know land and climate, how long it takes to integrate all the elements of a diversified farm. Small farmers can best build such a culture of competent local understanding and affection, as Wendell Berry has long argued. Given an even break, small private farmers have stronger incentives than either corporate or collective farms to work the land efficiently and to get the most from it—this has been abundantly proven. Given an ecologically sound economic framework and a secure footing, they will take good care of the land. This is why we need a tenacious private hold on most farms, preferably a hold that lasts for generations. The question is, How can we keep such people on the land?

We must turn to our common interest in farmland. Obviously, the common interest is not being well served by most private landowners today (including most farmers) because the present economic climate does not encourage deeply rooted small farms. Far from it: the present climate promotes unsustainable farming and the prompt sale of farmland to a higher and better use—gaudy palaces and tacky marts. This situation is likely to continue for some time. So how can the common interest in protecting farmland best be served today? Private farmers cannot meet it on their own—we need to help them along by imposing reasonable restraints. We must insist (as we increasingly do) that farming not cause ecological damage, such as polluting our water or allowing our soil to slip away at irreplaceable rates. We cannot legislate the best farming, but we can at least outlaw the worst. We can also subsidize the best practices, such as those that protect our watersheds. Such measures may increase our taxes or the price of food at the cash register, but that is as it should be. It will begin to swing the competitive advantage toward those farmers who best care for the land, where it belongs. Of course, this assumes that subsidies and regulations as they are actually written and administered can really promote good farming and the public interest in healthy food and land, instead of the interests of those best able to manipulate or evade the laws—a rather large assumption. But the alternative is to grant those same powerful interests a free market to work their will with the land, which is far worse.

Our strongest common interest lies in simply keeping farmland from being lost to development. We need to decide as particular communities that we are going to *need* local farmland in the not-too-distant future and that we *want* it even now, when the future need for it is still a matter of debate. If our goal is to keep family farmers on the land, preserving the possibility of the best farming, then our best tool is the conservation easement. By an easement, the community sequesters the right to develop the land, while the private owner remains otherwise in possession. In this way, both private and common interests are served. The farm family stays on the land but is taxed only on the residential value of its houselot (like any other family) and on the agricultural value of its farmland and improvements (like any other business). The exorbitant potential value of the farmland for residential development is absorbed by the community as a whole, thus expressing the common interest in preserving attractive, productive agricultural land close by. Either land trusts or municipalities can acquire and hold such easements. The difficulty, of course, is raising enough money to protect more than a token amount of the farmland within a community and doing so equitably. Our goal should not be modest. Our goal should be to keep anywhere from a quarter to half of the landscape (or even more depending on the terrain) in agricultural production—a monumental challenge at the urban cutting edge.

Clearly, the degree to which the common interest in protecting farmland needs to be actively asserted and the cost to the community of asserting that interest will vary greatly depending on the amount of development pressure. The need for common rights in farmland runs from slight in the hinterlands to severe at the city limit. Far away from the expanding metropolises, in happy regions that do not enjoy the mixed blessing of postcard scenic charm (but are only commonly beautiful), there may be little need to acquire conservation easements. Where demand for country estates isn't inflating the value of land, it will remain open on its own. At the other extreme, where very little open space remains there is much to be said for acquiring what is left for outright commons. Where people are thick there is an overwhelming need to secure common land for public enjoyment, whether as community farms and gardens or as parks. In between these poles lies the countryside where there is a common interest in keeping private farmers on land that the market now reckons is more valuable for dispersed housing.

The need for easements is strongest in places that are suburbanizing or enduring an influx of vacationers, snowbirds, or some other breed of wall-eyed urban refugee. The earlier the need is recognized, the more we can accomplish.

At the far end of the spectrum of intensity of land use lies wilderness, and close beside wilderness lies sustainably managed forest. Forestland belongs largely in common ownership. Ecological integrity and biodiversity provide little profit to any individual proprietor. Sharp accountants consider them economic liabilities. Therefore private owners have scant incentive to protect these values, beyond enlightened affection—which is admirable but doesn't last long in the market. Eventually, individuals who love the land for its own sake are forced to sell, can't resist selling, or die and leave the land to someone less affectionate. The *common* interest in ecological health is very strong, however. In the long run it is crucial that the ecological backbone of every community be permanently protected and placed under integrated local management. This might include several large, connected stretches of forest (or wetlands, savanna, prairie, as the case may be), amounting to several thousand acres in every town. Ideally, the community itself should own such conservation land and manage it for biodiversity, passive recreation, and sustainable wood and timber production.

Locally controlled common forests are desirable even in sparsely inhabited communities that are under little immediate threat of development. Common forests offer a secure basis for management at the landscape scale and place the means for local livelihood under local control. Privately owned forests are always vulnerable to acquisition and exploitation by outside interests, while state and federally owned public forests are too often effectively controlled by the same well-financed interlopers. By creating community forests we may also build the political constituency necessary to force better care of adjacent state and national forests. Locally owned common forests are certainly not immune from abuse, but I believe they provide the safest, most democratic

building blocks for the necessary continuous matrix of healthy natural ecosystems across the continent.

This is not to say that all forestland needs to be commonly
owned—there is plenty of forest out there for farmers and others
who wish to own private woodlots. But owners come and owners
go, while the urge to ravage forestland with either a feller-buncher
or a bulldozer never sleeps, except with one eye open. Let us aim for
an ecologically adequate chunk of permanently protected common
forest in each community. Let private woodlots attached to surrounding farms function as more flexible adjuncts that can move
from forest to agriculture and back over decades and centuries, as
the market and landowners determine. As a practical matter, of
course, the forest of most communities will probably be made up
largely of private forestland at best protected by easements, rather
than outright commons, for a long time to come. That is a good
place to start. Eventually, at least one-quarter and ideally one-half
of the land in every rural community should lie in common forest
and other natural ecosystems appropriate to the region.

There is another aspect of the landscape that deserves common
protection as much as its biological health: the most prominent hill-
tops, the shorelines of rivers and lakes, and other places of tran-
scendent beauty. These also ought to be commonly owned, and no
one ought to squat there, as Henry Thoreau insisted long ago.
"Think of a mountain top in the township ... only accessible
through private grounds," he wrote. "A temple as it were which
you cannot enter without trespassing—nay the temple itself private
property and standing in a man's cow yard—for such is commonly
the case. . . . That area should be left unappropriated for modesty
and reverence's sake—if only to suggest that the traveler who
climbs thither in a degree rises above himself, as well as his native
valley, and leaves some of his groveling habits behind."[1]

Yes, think of it. In Thoreau's time the trouble was that the private owner did not appreciate the beauty of the temple but saw it
with the eye of mere utility. In our time the owner appreciates the

beauty only too well but wants to engross it for himself—and not even be bothered to go outdoors and climb for it, but to guzzle it without rising from his easy chair. So much for modesty and reverence.

Like so many private appropriations of the commons, this self-aggrandizing impulse toward scenery is self-defeating in aggregate. What commands the view also occupies the view, and soon enough every noble outlook is defaced, and every private vista commands only similar eyesores. The picture window becomes an unerring mirror, reflecting inward. For the rest of us down below, a broad underview of all our superiors' groveling habits is fully revealed. Perhaps a single castle on a hill acquires a certain rustic charm after a thousand years or so, but when three and four and then a few dozen pop up, all the natural contours of the landscape are welted and scarred. People ought to have the common sense to build lower on the slopes, but of course as detached individuals they don't, vying instead for the peaks. Protecting the charming face of a community is just as important as protecting its ecological bones, and fortunately the strongest skeleton often underlies the most handsome features. Those working to build common forests should secure the scenic ridgetops and riparian areas first, before the speculators ride into the valley with their fat wallets and narrow vision, scanning the horizon for prospects. The best lookouts in Weston belong to our commons, and we can all climb thither in peace, without glancing uneasily over our shoulder for the puffed-up proprietor of the day.

That is a vision of how rural communities might be reinhabited and kept whole. With a little foresight, development could be confined to a small percentage of the land in a central village and outlying hamlets and clusters, while a mixture of private farms protected by easements and forests owned in common could possess the balance of the landscape in perpetuity. Needless to say, in the real world acquiring such easements and commons will be painfully costly and slow. We will inevitably fall far short of the ideal and for a long time lose much more than we save. In many places it is already too late to achieve anything like an ideal reset-tlement, but we must keep working to do what we can along these lines—every small victory gives us more to work with for the future, more land with which to change minds. Someday, resettling rural areas in such a way may become a necessity, and the more we know about how to do it, the better. We must be like the belea-guered forest a century ago, holding on to all we can and sending forth our ideas like seeds, encased in enough accumulated experi-ence to nurture them. A few will fall on fertile ground.

At the moment, we are in an unequal race against unbridled development that is rapidly harrowing the landscape, with seem-ingly endless ready cash to spend. One may well wonder where the countervailing funds for commons will be found. In some cases public-spirited landowners may donate easements for love because they wish to ensure their land remains undeveloped and to reduce their property and inheritance taxes. But more often common lands and easements will have to be purchased by the community. Local citizens will have to be democratically convinced that it is in their interest to tax themselves for this purpose, as has happened in Weston. This may save a few key parcels of land, but that will only get us so far.

Fortunately, the force of invading development can sometimes be turned against itself—the art of self-defense judo for threatened land, if you like. One approach is to impose a tax of 1 or 2 percent on every sale of real estate, the proceeds going into a land bank to purchase conservation land. This is really just a property tax in a palatable and appropriate form to address a specific need. In this way a small part of the rising price that newcomers are willing to pay for property can be used to protect open space for public use. Maryland and Vermont have enacted such taxes. Massachusetts has successful land banks on Martha's Vineyard and Nantucket, and now a wave of towns is pushing for such measures in spite of stiff opposition from the real estate lobby. By floating large bonds against an assured source of revenue stretching into the future, communities can protect a substantial amount of land before it is too late.

Another way to limit sprawl is to promote partial, clustered development, in place of cookie-cutter build-out. In this way part of a threatened parcel can be densely developed, thus protecting the remainder. Clustered development (or conservation subdivision, as it is sometimes called) appears to be the most effective tool we have to save land, and town planning boards should be going all out to encourage it. Sometimes complex deals can be put together that include the clustering of a reduced number of homes, a conservation easement over most of the parcel, and the outright purchase of the most attractive part of the land for commons. Clustered developments can be ecologically well designed themselves, and well integrated with the protected open space surrounding them.

All this marks a radical departure in how we own land, no matter how it is couched. Transferring anything like one-half of the forest from private hands to commons and holding protective easements over a substantial portion of the remaining private farmland would of course mark a fundamental transformation of American landownership. It surely will not happen overnight. But if it does not happen over time it is hard to envision how we will be able to protect the farmland and forest we need to live sustainably, let alone enjoyably. A revolution in landownership is needed. But I want to repeat that I am by no means calling for the forced expropriation of private property, for a vanguard of the eco-tariat to impose collective farms, cattle-prod in one hand and Kalashnikov in the other. I am not demanding that corporate middle-managers surrender their backyards, shoulder their grubhoes, and march to the common fields for reeducation. I envision towns in which private and common property rights flourish side by side and often overlap, drawing on the strengths of each. This is in fact what we are starting to see in towns such as Weston—hardly a hotbed of socialism.

Thoughtful agrarians are only selectively nostalgic. The point is not to return to the Middle Ages but to adapt the principle of the commons to the modern world. In the medieval world of resource scarcity, traditional systems brought together common and private control over different kinds of land in useful ways. Commons systems were designed not to maximize production but to optimize ecological security. This was a conservative approach to landownership, and I think we will need something like it again as we approach a new set of ecological limits in the world. Lop off the overlords, insist on protecting ecological values on common forestland, acquire a measure of common protection for private farmland, and we will have a very effective and appropriate mix of landownership for the new millennium.

But aren't commons by their very nature hopelessly inefficient, if not prone to tragic abuse? I need to say a word here about the widely held but mistaken belief that commons are invariably wrecked by the commoners themselves, whereas private owners tend to take good care of their land—the so-called tragedy of the commons, in Garrett Hardin's unfortunate and misleading phrase. Hardin coined the term as a metaphor for the unlimited right to bear children in a crowded world, but it has been heartily endorsed in its literal sense by the property rights movement. The theory holds that because an individual has a strong personal incentive to cheat on the commons and a much weaker incentive to care for it, any commonly owned property is bound to be overexploited and thus degraded. This profoundly underestimates human nature—not private virtue, but collective jealousy. In actual historical fact, virtually all commons were closely guarded against individuals' cheating by the watchful eyes of their neighbors, backed up, when necessary, by the law. Commons were also vigorously defended against outside intruders. The term *commons* includes not only a common resource, but a mechanism for community control. There is no such thing as a free commons: if it's free, then it's not a commons. Commons have bounds and rules that limit access. Commons, like democracy, must always be defended. Given that, the system worked remarkably well for long periods of time in many parts of the world.

The corollary of the tragedy of the commons argument holds that private ownership ensures better treatment of land because the owner has a strong incentive to care for it. This assertion is so childishly simplistic as to be disingenuous. At best, it profoundly overestimates private virtue and resistance to the temptations of the market. A powerful incentive to care for the homestead may exist where a family hopes to live in one place for generations, and that is why I favor private ownership of farmland—once the urge to liquidate the ancestral land for profit has been eased. But in most cases in our culture, private ownership is brief. It is simply a license to exploit the land for all it is worth and then to cast it aside while reinvesting the profits elsewhere. The owner is not obliged to live on the property after finishing with it and often never lives there at all. To understand that private ownership is no panacea, one need look only at the cut-and-run practices of nineteenth-century timber barons (or at the similar behavior of some corporate forest owners today, for that matter), who let the land become a public charge after they had stripped it. Private ownership certainly does have a productive role to play if limitations can be placed on the rights of ownership and if the market system in which it is embedded can be modified to give the owner ecologically responsible signals.

Ideological quarrels that pit common against private property as ideal types are pointless. Each has its place. Common ownership cannot guarantee good management, but it is a sensible choice for that part of the landscape where the highest priority is restraint, rather than productivity. Private ownership is generally more productive but ecologically more risky. Even in long-settled New England, with its strong heritage of devotion to place, privatization of common forest and grazing land led in time to devastated woodlands and rundown pastures. Once the market system was firmly in the saddle by the nineteenth century, private farmers (some of

whom represented the proverbial seventh generation) often chose
to cash in their ancestral legacy by "skinning the land" rather than
husband it any longer. Common ownership may not often lead to
the most efficient production of wealth, but it is an appropriate
means to safeguard long-term ecological health. The tragedy of the
commons seldom lies in common ownership itself. The tragedy
usually lies in the expropriation of common resources for unre-
stricted private gain. Privatization is seldom the solution to the
tragedy of the commons—more often privatization *is* the tragedy of
the commons. The solution is to reclaim and strengthen the commons.

The commons is not a utopian ideal of ownership of everything
by everybody. There are things like the Earth's ozone layer that can
be considered a global commons and need to be protected by international agreements. Our national and state forests are sometimes
called commons, but, being so large, they must be controlled by
bureaucracies; they become battlegrounds of competing interests.
The kinds of commons I am discussing here are small, and the
bounds are not far from home. They can be perambulated in a day.
One of the crucial ideas in a commons system is that the land be
controlled by the *local* community, by people who know it well and
must live with the consequences of their actions upon it and with
each other. This requires a community whose residents know how
to speak to each other—which is perhaps why commons have reap-
peared in New England, with its town meeting form of direct
democracy.

A commons also requires commoners who are productively
engaged with the land, beyond weekend birding or mountain bik-
ing (I speak as an avid birder and biker, of course). For this reason
a commons system and a revived local economy must go hand in
hand. Such an economy will require broad participation, if that
part of the community's land which is common is to be governed by
residents who fully appreciate its value. We are not likely, however,
to see the rise of a new agrarian class of Jeffersonian small family

farmers any time soon, even though movement in that direction would be healthy and desirable. One can still sanely dream of such a revival taking place someday in parts of the Midwest that have been emptied, but on the densely settled eastern seaboard and in other suburbanizing regions that would be a flight of purest fancy. What we can envision here, I believe, is a broader, common agrarianism in every community that includes a few dozen full-time farmers and woodsworkers, a wider range of part-time farmers, and one community farming and forestry program that involves most residents with the land in some meaningful way. In this way, we can make caring for the land a normal part of growing up in rural, suburban, and even urban communities—a common experience, if you will. This will provide our commons with the necessary commoners.

To have a healthy relation with the land, we need places that are held together by diversified local economies. This is true whether industrial society is entering an era of continued prosperity and expansion or one of tightening limits, scarcity, and decline—or one gradually being overtaken by the other, as now seems most likely to me. In this new century, our species will face enormous ecological challenges. We must make the transition to an energy system that no longer relies heavily on fossil fuel, meanwhile feeding, clothing, and housing some ten billion people without exhausting and destroying our soil and water. Next, we must do this without woefully diminishing the complex biosphere composed of millions of fellow-species that makes our planet habitable and beautiful. How are we to meet these human and biological imperatives at the same time, from the same ground? The population of the United States is expected to increase 50 percent by the middle of this century to nearly four hundred million, and we are people who are in the habit of consuming exorbitant quantities of resources. Clearly, some changes in the way we use land are in store.

I am unable to predict how well we will do at meeting these challenges. In either case, whether we manage to spread material prosperity and comforts widely among the members of our species or suffer such setbacks that most of us barely scrape by at the level of survival, we will benefit from a new, ecologically sound, common agrarianism. It may be forced upon us by necessity or we may find it to be a healthy and rewarding way for prosperous people to live. The sooner we come to it by choice rather than dire need, the better.

Chapter 14

The Common Life

Scott Russell Sanders

The agrarian persuasion, as Wendell Berry explains, is fundamentally a matter of living responsibly at home. With his family, Scott Russell Sanders has sought to do just that in his chosen community of Bloomington, Indiana. Over his years there, Sanders has come to know the surrounding countryside, the plants, animals, rivers, and soils that make up and share his home. He has also come to know the people and has sought as best he can to join with them in ways that sustain their collective well-being. In his nonfiction essays, chiefly autobiographical, Sanders reports on his progress at that work. He undertakes it with an infectious zest for the fundamentals of life—including baking bread and enjoying the company of neighborhood children, which he describes here. He relates, "The elements of my kitchen scene—loving company, neighborliness, inherited knowledge and good work, shared purpose, sensual delight, and union with the creation—sum up for me what is vital in community."

This essay, an exceptional portrait of what ought to be ordinary, comes from Sanders' collection Writing from the Center, *a book written from both the center of the country and the center of his life. Sanders' life and writing make plain the possibilities for the agrarian persuasion to thrive in a city when people care for one another, for their communal structures, and for the land.*

One delicious afternoon while my daughter Eva was home from college for spring vacation, she invited two neighbor girls to help her make bread. The girls are sisters, five-year-old Alexandra and ten-year-old Rachel, both frolicky, with eager dark eyes and shin-

ing faces. They live just down the street from us here in Bloomington, Indiana, and whenever they see me pass by, on bicycle or on
foot, they ask about Eva, whom they adore.

I was in the yard that afternoon mulching flower beds with
compost, and I could hear the girls chattering as Eva led them up
the sidewalk to our door. I had plenty of other chores to do in the
yard, where every living thing was urgent with April. But how
could I stay outside, when so much beauty and laughter and spunk
were gathered in the kitchen?

I kept looking in on the cooks, until Eva finally asked, "Daddy,
you wouldn't like to knead some dough, would you?"

"I'd love to," I said. "You sure there's room for me?"

"There's room," Eva replied, "but you'll have to wash in the
basement."

Hands washed, I took my place at the counter beside Rachel
and Alexandra, who perched on a stool I had made for Eva when
she was a toddler. Eva had still needed that stool when she learned
to make bread on this counter; and my son, now six feet tall, had
balanced there as well for his own first lessons in cooking. I never
needed the stool, but I needed the same teacher—my wife, Ruth, a
woman with eloquent fingers.

Our kitchen is small; Ruth and I share that cramped space by
moving in a kind of dance we have been practicing for years. When
we bump one another, it is usually for the pleasure of bumping. But
Eva and the girls and I jostled like birds too numerous for a nest.
We spattered flour everywhere. We told stories. We joked. All the
while I bobbed on a current of bliss, delighting in the feel of live
dough beneath my fingers, the smell of yeast, the piping of child-
voices so much like the birdsong cascading through our open windows, the prospect of whole-wheat loaves hot from the oven.

An artist might paint this kitchen scene in pastels for a poster,
with a tender motto below, as evidence that all is right with the
world. All is manifestly *not* right with the world. The world, most
of us would agree, is a mess: rife with murder and mayhem, abuse

of land, extinction of species, lying and theft and greed. There are days when I can see nothing but a spectacle of cruelty and waste, and the weight of dismay pins me to my chair. On such days I need a boost merely to get up, uncurl my fists, and go about my work. The needed strength may come from family, from neighbors, from a friend's greeting in the mail, from the forked leaves of larkspur breaking ground, from rainstorms and music and wind, from the lines of a handmade table or the lines in a well-worn book, from the taste of an apple or the brash trill of finches in our backyard trees. Strength also comes to me from memories of times when I have felt a deep and complex joy, a sense of being exactly where I should be and doing exactly what I should do, as I felt on that breadmaking afternoon.

I wish to reflect on the sources of that joy, that sense of being utterly in place, because I suspect they are the sources of all that I find authentic in my life. So much in life seems to me unauthentic, I cannot afford to let the genuine passages slip by without considering what makes them ring true. It is as though I spend my days wandering about, chasing false scents, lost, and then occasionally, for a few ticks of the heart, I stumble onto the path. While making bread with my daughter and her two young friends, I was on the path. So I recall that time now as a way of keeping company with Eva, who has gone back to college, but also as a way of discovering in our common life a reservoir of power and hope.

❈　❈　❈

What is so powerful, so encouraging, in that kitchen scene? To begin with, I love my three fellow cooks; I relish every tilt of their heads and turn of their voices. In their presence I feel more alive and alert, as if the rust had been knocked off my nerves. The armor of self dissolves, ego relaxes its grip, and I am simply there, on the breeze of the moment.

Rachel and Alexandra belong to the Abed family, with whom

we often share food and talk and festivities. We turn to the Abeds
for advice, for starts of plants, for cheer, and they likewise turn to
us. Not long ago they received troubling news that may force them
to move away, and we have been sharing in their distress. So the
Abed girls brought into our kitchen a history of neighborliness, a
history all the more valuable because it might soon come to an end.

The girls also brought a readiness to learn what Eva had to
teach. Eva, as I mentioned, had learned from Ruth how to make
bread, and Ruth had learned from a Canadian friend, and our
friend had learned from her grandmother. As Rachel and Alexandra shoved their hands into the dough, I could imagine the rope of
knowledge stretching back and back through generations, to folks
who ground their grain with stones and did their baking in wood
stoves or fireplaces or in pits of glowing coals.

If you have made yeast bread, you know how at first the dough
clings to your fingers, and then gradually, as you knead in more
flour, it begins to pull away and take on a life of its own, becoming
at least as resilient as a plump belly. If you have not made yeast
bread, no amount of hearing or reading about it will give you that
knowledge, because you have to learn through your body, through
fingers and wrists and aching forearms, through shoulders and
back. Much of what we know comes to us that way, passed on from
person to person, age after age, surviving in muscle and bone. I
learned from my mother how to transplant a seedling, how to sew
on a button; I learned from my father how to saw a board square,
how to curry a horse, how to change the oil in a car. The pleasure I
take in sawing or currying, in planting or sewing, even in changing
oil, like my pleasure in making bread, is bound up with the affection I feel for my teachers and the respect I feel for the long, slow
accumulation of knowledge that informs our simplest acts.

Those simple acts come down to us because they serve real
needs. You plant a tree or sweep a floor or rock a baby without asking the point of your labors. You patch the roof to stop a leak, patch
a sweater to keep from having to throw it out. You pluck the banjo

because it tickles your ears and rouses Grandpa to dance. None of us can live entirely by such meaningful acts; our jobs, if nothing else, often push us through empty motions. But unless at least some of what we do has a transparent purpose, answering not merely to duty or fashion but to actual needs, then the heart has gone out of our work. What art could be more plainly valuable than cooking? The reason for baking bread is as palpable as your tongue. After our loaves were finished, Eva and I delivered two of them to the Abeds, who showed in their faces a perfect understanding of the good of bread.

When I compare the dough to a plump belly, I hear the sexual overtones, of course. By making the comparison, I do not wish to say, with Freud, that every sensual act is a surrogate for sex; on the contrary, I believe sex comes closer to being a stand-in, rather brazen and obvious, like a ham actor pretending to be the whole show, when it is only one player in the drama of our sensual life. That life flows through us constantly, so long as we do not shut ourselves off. The sound of birds and the smell of April dirt and the brush of wind through the open door were all ingredients in the bread we baked.

Before baking, the yeast was alive, dozing in the refrigerator. Scooped out of its jar, stirred into warm water, fed on sugar, it soon bubbled out gas to leaven the loaves. You have to make sure the water for the yeast, like milk for a baby, is neither too hot nor too cold, and so, as for a baby's bottle, I test the temperature on my wrist. The flour, too, had been alive not long before as wheat thriving in sun and rain. Our nourishment is borrowed life. You need not be a Christian to feel, in a bite of bread, a sense of communion with the energy that courses through all things. The lump in your mouth is a chunk of earth; there is nothing else to eat. In our house we say grace before meals, to remind ourselves of that gift and that dependence.

The elements of my kitchen scene—loving company, neighborliness, inherited knowledge and good work, shared purpose, sensual delight, and union with the creation—sum up for me what is vital in community. Here is the spring of hope I have been led to by my trail of bread. In our common life we may find the strength not merely to carry on in face of the world's bad news, but to resist cruelty and waste. I speak of it as common because it is ordinary, because we make it together, because it binds us through time to the rest of humanity and through our bodies to the rest of nature. By honoring this common life, nurturing it, carrying it steadily in mind, we might renew our households and neighborhoods and cities, and in doing so might redeem ourselves from the bleakness of private lives spent in frenzied pursuit of sensation and wealth.

❉　❉　❉

Ever since the eclipse of our native cultures, the dominant American view has been more nearly the opposite: that we should cultivate the self rather than the community; that we should look to the individual as the source of hope and the center of value, while expecting hindrance and harm from society.

What other view could have emerged from our history? The first Europeans to reach America were daredevils and treasure seekers, as were most of those who mapped the interior. Many colonists were renegades of one stripe or another, some of them religious nonconformists, some political rebels, more than a few of them fugitives from the law. The trappers, hunters, traders, and freebooters who pushed the frontier westward seldom recognized any authority beyond the reach of their own hands. Coast to coast, our land has been settled and our cities have been filled by generations of immigrants more intent on leaving behind old tyrannies than on seeking new social bonds.

Our government was forged in rebellion against alien control. Our economy was founded on the sanctity of private property, and thus our corporations have taken on a sacred immunity through being defined under the law as persons. Our criminal justice system is so careful to protect the rights of individuals that it may require years to convict a bank robber who killed a bystander in front of a crowd, or a bank official who left a trial of embezzlement as wide as the Mississippi.

Our religion has been marked by an evangelical Protestantism that emphasizes personal salvation rather than social redemption. To "Get Right with God," as signs along the roads here in the Midwest gravely recommend, does not mean to reconcile your fellow citizens to the divine order, but to make a separate peace, to look after the eternal future of your own singular soul. True, we have a remarkable history of communal experiments, most of them religiously inspired—from Plymouth Colony, through the Shaker villages, Robert Owen's New Harmony, the settlements at Oneida, Amana, and countless other places, to the communes in our own day. But these are generally known to us, if they are known at all, as utopian failures.

For much of the present century, Americans have been fighting various forms of collectivism—senile empires during World War I, then Nazism, communism, and now fundamentalist theocracies—and these wars, the shouting kind as well as the shooting kind, have only strengthened our commitment to individualism. We have understood freedom for the most part negatively rather than positively, as release from constraints rather than as the condition for making a decent life in common. Hands off, we say; give me elbow room; good fences make good neighbors; my home is my castle; don't tread on me. I'm looking out for number one, we say; I'm doing my own thing. We have a Bill of Rights, which protects each of us from a bullying society, but no Bill of Responsibilities, which would oblige us to answer the needs of others.

Even where America's founding documents clearly address the public good, they have often been turned to private ends. Consider just one notorious example, the Second Amendment to the Constitution:

> A well regulated Militia, being necessary to the secu-
> rity of a free State, the right of the people to keep and
> bear Arms, shall not be infringed.

It would be difficult to say more plainly that arms are to be kept
for the sake of a militia, and a militia is to be kept for defense of the
country. In our day, a reasonable person might judge that the Pentagon deploys quite enough weapons, without requiring any sup-
plement from household arsenals. Yet this lucid passage has been
construed to justify a domestic arms race, until we now have in
America more gun shops than gas stations, we have nearly as many
handguns as hands, and we have concentrated enough firepower in
the average city to carry on a war—which is roughly what, in some
cities, is going on. Thus, by reading the Second Amendment
through the lens of our obsessive individualism, we have turned a
provision for public safety into a guarantee of public danger.

Observe how zealously we have carved up our cities and paved
our land and polluted our air and burned up most of the earth's
petroleum within a single generation—all for the sake of the automobile, a symbol of personal autonomy even more potent than the
gun. There is a contemptuous ring to the word "mass" in mass
transportation, as if the only alternative to private cars were cattle
cars. Motorcycles and snowmobiles and three-wheelers fill our public lands with the din of engines and tear up the terrain, yet any
effort to restrict their use is denounced as an infringement of individual rights. Millions of motorists exercise those rights by hurling
the husks of their pleasures onto the roadside, boxes and bottles and
bags. Ravines and ditches in my part of the country are crammed
with rusty cars and refrigerators, burst couches and stricken TVs,
which their former owners would not bother to haul to the dump.
Meanwhile, advertisers sell us everything from jeeps to jeans as
tokens of freedom, and we are so infatuated with the sovereign self
that we fall for the spiel, as if by purchasing one of a million iden-
tical products we could distinguish ourselves from the herd.

The cult of the individual shows up everywhere in American lore, which celebrates drifters, rebels, and loners, while pitying or reviling the pillars of the community. The backwoods explorer like Daniel Boone, the riverboat rowdy like Mike Fink, the lumberjack, the prospector, the rambler and gambler, the daring crook like Jesse James and the resourceful killer like Billy the Kid, along with countless lonesome cowboys, all wander, unattached, through the great spaces of our imagination. When society begins to close in, making demands and asking questions, our heroes hit the road. Like Huckleberry Finn, they are forever lighting out for the Territory, where nobody will tell them what to do. Huck Finn ran away from what he called civilization in order to leave behind the wickedness of slavery, and who can blame him, but he was also running away from church and school and neighbors, from aunts who made him wash before meals, from girls who cramped his style, from chores, from gossip, from the whole nuisance of living alongside other people.

In our literature, when community enters at all, it is likely to appear as a conspiracy against the free soul of a hero or heroine. Recall how restless Natty Bumppo becomes whenever Cooper drags him in from the woods to a settlement. Remember how strenuously Emerson preaches against conforming to society and in favor of self-reliance, how earnestly Hawthorne warns us about the tyranny of those Puritan villages. Think of Thoreau running errands in Concord, rushing in the front door of a house and out the back, then home to his cabin in the woods, never pausing, lest he be caught in the snares of the town. Think of the revulsion Edna Pontellier feels toward the Creole society of New Orleans in Kate Chopin's *The Awakening*. Think of Willa Cather's or Toni Morrison's or James Baldwin's high-spirited women and men who can only thrive by fleeing their home communities. Think of Spoon River, Winesburg, Gopher Prairie, Zenith, all those oppressive fictional places, the backward hamlets and stifling suburbs and heartless cities that are fit only for drones and drudges and mindless Babbitts.

In *The Invasion of the Body Snatchers,* a film from my childhood that still disturbs my dreams, an alien life form takes over one person after another in a small town, merging them into a single creature with a single will, until just one freethinking individual remains, and even he is clearly doomed. Along with dozens of other invasion tales, the film was a warning against communism, I suppose, but it was also a caution against the perils of belonging, of losing your one sweet self in the group, and thus it projected a fear as old as America.

Of course, you can find American books and films that speak as passionately for the virtues of our life together as for the virtues of our lives apart. To mention only a few novels from the past decade, I think of Gloria Naylor's *Mama Day,* Wendell Berry's *A Place on Earth,* Ursula Le Guin's *Always Coming Home,* and Ernest Gaines's *A Gathering of Old Men.* But they represent a minority opinion. The majority opinion fills bestseller lists and cinema screens and billboards with isolated, alienated, rebellious figures who are too potent or sensitive for membership in any group.

❖   ❖   ❖

I have been shaped by this history, and I, too, am uneasy about groups, especially large ones, above all those that are glued together by hatred, those that use a color of skin or a cut of clothes for admission tickets, and those that wrap themselves in scriptures or flags. I have felt a chill from blundering into company where I was not wanted. I have known women and men who were scorned because they refused to fit the molds their neighbors had prepared for them. I have seen Klansmen parading in white hoods, their crosses burning on front lawns. I have seen a gang work its way through a subway car, picking on the old, the young, the weak. Through film I have watched the Nuremberg rallies, watched policemen bashing demonstrators in Chicago, missiles parading in Red Square, tanks crushing dissidents in Tiananmen Square. Like everyone born since World War II, I have grown up on television images of atrocities carried out, at home and abroad, with the blessing of governments or revolutionary armies or charismatic thugs.

In valuing community, therefore, I do not mean to approve of any and every association of people. Humans are drawn together by a variety of motives, some of them worthy, some of them ugly. Anyone who has spent recess on a school playground knows the terror of mob rule. Anyone who has lived through a war knows that mobs may pretend to speak for an entire nation. I recently saw, for the first time in a long while, a bumper sticker that recalled for me the angriest days of the Vietnam War: AMERICA—LOVE IT OR LEAVE IT. What loving America seemed to mean, for those who brandished that slogan back around 1970, was the approval of everything the government said or the army did in our name. "All those who seek to destroy the liberty of a democratic nation ought to know," Alexis de Tocqueville observed in *Democracy in America,* "that war is the surest and the shortest means to accomplish it." As a conscientious objector, with a sister who studied near my home in Ohio on a campus where National Guardsmen killed several protesters, I felt the force of despotism in that slogan.

Rather than give in to despotism, some of my friends went to jail and others into exile. My wife and I considered staying in England, where we had been studying for several years and where I had been offered a job. But instead we chose to come home, here to the Midwest where we had grown up, to work for change. In our idealism, we might have rephrased that bumper sticker to read: AMERICA—LOVE IT AND REDEEM IT. For us, loving America had little to do with politicians and even less with soldiers, but very much to do with what I have been calling the common life: useful work, ordinary sights, family, neighbors, ancestors, our fellow creatures, and the rivers and woods and fields that make up our mutual home.

During the more than twenty years since returning to America, I have had some of the idealism knocked out of me, but I still believe that loving your country or city or neighborhood may require you to resist, to call for change, to speak on behalf of what you believe in, especially if what you believe in has been neglected.

What we have too often neglected, in our revulsion against tyranny and our worship of the individual, is the common good. The results of that neglect are visible in the decay of our cities, the despoiling of our land, the fouling of our rivers and air, the haphazard commercial sprawl along our highways, the gluttonous feeding at the public trough, the mortgaging of our children's and grandchildren's future through our refusal to pay for current consumption. Only a people addicted to private pleasure would allow themselves to be defined as consumers—rather than conservers or restorers—of the earth's abundance.

In spite of the comforting assurances, from Adam Smith onward, that the unfettered pursuit of private wealth should result in unlimited public good, we all know that to be mostly a lie. If we needed reminders of how great that lie is, we could look at the savings and loan industry, where billions of dollars were stolen from small investors by rich managers who yearned to be richer; we could look at the Pentagon, where contracts are routinely encrusted with graft, or at Wall Street, where millionaires finagle to become billionaires through insider trading; we could look at our national forests, where logging companies buy timber for less than the cost to taxpayers of harvesting it; or we could look at our suburbs, where palaces multiply, while downtown more and more people are sleeping in cardboard boxes. Wealth does not precipitate like dew from the air; it comes out of the earth and from the labor of many hands. When a few hands hold onto great wealth, using it only for personal ease and display, that is a betrayal of the common life, the sole source of riches.

❊   ❊   ❊

Fortunately, while our tradition is heavily tilted in favor of private life, we also inherit a tradition of caring for the community.

Although Tocqueville found much to fear and quite a bit to despise in this raw democracy, he praised Americans for having "carried to the highest perfection the art of pursuing in common the object of their common desires." Writing of what he had seen in the 1830s, Tocqueville judged Americans to be avaricious, self-serving, and aggressive; but he was also amazed by our eagerness to form clubs, to raise barns or town halls, to join together in one cause or another: "In no country in the world, do the citizens make such exertions for the common weal. I know of no people who have established schools so numerous and efficacious, places of public worship better suited to the wants of the inhabitants, or roads kept in better repair."

Today we might revise his estimate of our schools or roads, but we can still see all around us the fruits of that concern for the common weal—the libraries, museums, courthouses, hospitals, orphanages, universities, parks, on and on. Born as most of us are into places where such amenities already exist, we may take them for granted; but they would not be there for us to use had our forebears not cooperated in building them. No matter where we live, our home places have also benefited from the Granges and unions, the volunteer fire brigades, the art guilds and garden clubs, the charities, food kitchens, homeless shelters, soccer and baseball teams, the Scouts and 4-H, the Girls and Boys Clubs, the Lions and Elks and Rotarians, the countless gatherings of people who saw a need and responded to it.

This history of local care hardly ever makes it into our literature, for it is less glamorous than rebellion, yet it is a crucial part of our heritage. Any of us could cite examples of people who dug in and joined with others to make our home places better places. Women and men who invest themselves in their communities, fighting for good schools or green spaces, paying attention to where they are, seem to me as worthy of celebration as those adventurous loners who keep drifting on, prospecting for pleasure.

A few days after our breadmaking, Eva and I went to a concert in Bloomington's newly opened arts center. The old limestone building had once been the town hall, then a fire station and jail, then for several years an abandoned shell. Volunteers bought the building from the city for a dollar, then renovated it with materials, labor, and money donated by local people. Now we have a handsome facility that is in constant use for pottery classes, theater productions, puppet shows, art exhibits, poetry readings, and every manner of musical event.

The music Eva and I heard was *Hymnody of Earth,* for hammer dulcimer, percussion, and children's choir. Composed by our next-door neighbor, Malcolm Dalglish, and featuring lyrics by our Ohio Valley neighbor, Wendell Berry, it was performed that night by Malcolm, percussionist Glen Velez, and the Bloomington Youth Chorus. As I sat there with Eva in a sellout crowd—about a third of whom I knew by name, another third by face—I listened to music that had been elaborated within earshot of my house, and I heard my friend play his instrument, and I watched those children's faces shining with the colors of the human spectrum, and I felt the restored building clasping us all like the cupped hands of our community. I knew once more that I was in the right place, a place created and filled and inspired by our lives together.

❖   ❖   ❖

A woman who recently moved from Los Angeles to Bloomington told me that she would not be able to stay here long, because she was already beginning to recognize people in the grocery stores, on the sidewalks, in the library. Being surrounded by familiar faces made her nervous, after years in a city where she could range about anonymously. Every traveler knows the sense of liberation that comes from journeying to a place where nobody expects anything of you. Everyone who has gone to college knows the exhilaration of slipping away from the watchful eyes of Mom and Dad. We all need seasons of withdrawal from responsibility. But if we make a career of being unaccountable, we have lost something essential to our humanity, and we may well become a burden or a threat to those around us. A community can support a number of people who are just passing through, or who care about no one's needs but their own; the greater the proportion of such people, however, the move vulnerable the community, until eventually it breaks down. That is true on any scale, from a household to a planet.

The words *community, communion,* and *communicate* all derive from *common,* and the two syllables of *common* grow from separate roots, the first meaning "together" or "next to," the second having to do with barter or exchange. Embodied in that word is a sense of our shared life as one of giving and receiving—music, touch, ideas, recipes, stories, medicine, tools, the whole range of artifacts and talents. After twenty-five years with Ruth, that is how I have come to understand marriage, as a constant exchange of labor and love. We do not calculate who gives how much; if we had to, the marriage would be in trouble. Looking outward from this community of two, I see my life embedded in ever larger exchanges—those of family and friendship, neighborhood and city, countryside and country—and on every scale there is giving and receiving, calling and answering.

Many people shy away from community out of a fear that it may become suffocating, confining, even vicious; and of course it may, if it grows rigid or exclusive. A healthy community is dynamic, stirred up by the energies of those who already belong, open to new members and fresh influences, kept in motion by the constant bartering of gifts. It is fashionable just now to speak of this open quality as "tolerance," but that word sounds too grudging to me—as though, to avoid strife, we must grit our teeth and ignore whatever is strange to us. The community I desire is not grudging; it is exuberant, joyful, grounded in affection, pleasure, and mutual aid. Such a community arises not from duty or money but from the free interchange of people who share a place, share work and food, sorrows and hope. Taking part in the common life means dwelling in a web of relationships, the many threads tugging at you while also holding you upright.

I have told elsewhere the story of a man who lived in the Ohio township where I grew up, a builder who refused to join the volunteer fire department. Why should he join, when his house was brick, properly wired, fitted out with new appliances? Well, one day that house caught fire. The wife dialed the emergency number, the siren wailed, and pretty soon the volunteer firemen, my father among them, showed up with the pumper truck. But they held back on the hoses, asking the builder if he still saw no reason to join, and the builder said he could see a pretty good reason to join right there and then, and the volunteers let the water loose.

I have also told before the story of a family from that township whose house burned down. The fire had been started accidentally by the father, who came home drunk and fell asleep smoking on the couch. While the place was still ablaze, the man took off, abandoning his wife and several young children. The local people sheltered the family, then built them a new house. This was a poor township. But nobody thought to call in the government or apply to a foundation. These were neighbors in a fix, and so you helped them, just as you would harvest corn for an ailing farmer or pull a flailing child from the creek or put your arm around a weeping friend.

I am not harking back to some idyllic past, like the one embalmed in the *Saturday Evening Post* covers by Norman Rockwell or the prints of Currier and Ives. The past was never golden. As a people, we still need to unlearn some of the bad habits formed during the long period of settlement. One good habit we might reclaim, however, is that of looking after those who live nearby. For much of our history, neighbors have kept one another going, kept one another sane. Still today, in town and country, in apartment buildings and barrios, even in suburban estates, you are certain to lead a narrower life without the steady presence of neighbors. It is neither quaint nor sentimental to advocate neighborliness; it is far more sentimental to suggest that we can do without such mutual aid.

Even Emerson, preaching self-reliance, knew the necessity of neighbors. He lived in a village, gave and received help, and delivered his essays as lectures for fellow citizens whom he hoped to sway. He could have left his ideas in his journals, where they first took shape, but he knew those ideas would only have effect when they were shared. I like to think he would have agreed with the Lakota shaman, Black Elk, that "a man who has a vision is not able to use the power of it until after he has performed the vision on earth for the people to see." If you visit Emerson's house in Concord, you will find leather buckets hanging near the door, for he belonged to the village fire brigade, and even in the seclusion of his study, in the depths of thought, he kept his ears open for the alarm bell.

We should not have to wait until our houses are burning before we see the wisdom of facing our local needs by joining in common work. We should not have to wait until gunfire breaks out in our schools, rashes break out on our skin, dead fish float in our streams, or beggars sleep on our streets before we act on behalf of the community. On a crowded planet, we had better learn how to live well together, or we will live miserably apart.

❈ ❈ ❈

In cultural politics these days there is much talk of diversity and difference. This is all to the good, insofar as it encourages us to welcome the many distinctive traditions and visions that have flowed into America from around the world. But if, while respecting how we differ, we do not also recognize how much we have in common, we will have climbed out of the melting pot into the fire. Every day's newspaper brings word of people suffering and dying in the name of one distinction or another. We have never been slow to notice differences—of accent, race, dress, habits. If we merely change how those differences are valued, celebrating what had formerly been despised or despising what had been celebrated, we continue to define ourselves and one another in the old divisive ways.

Ethnic labels are especially dangerous, for, while we pin them on as badges of pride, we may have difficulty taking them off when others decide to use them as targets. The larger the group identified by a label, the less likely it is to be a genuine community. Haste or laziness may lead us to speak of blacks and whites, of Christians and Muslims and Jews, of Armenians and Mexicans, yet the common life transcends such categories. Sharing a national anthem, a religion, or a skin color may be grounds for holding rallies or waging war, but community is more intimate than nationality, more subtle than race or creed, arising not from abstract qualities but from the daily give-and-take among particular people in a particular place.

It is also dangerous to separate a concern for human diversity from a concern for natural diversity. Since Europeans arrived in North America, we have been drawing recklessly on beaver and bison, trees and topsoil, petroleum, coal, iron and copper ore, clean air and water. Many of the achievements on which we pride ourselves are the result not of our supposed virtues but of this plundered bounty. We do not have another continent to use up; unless we learn to inhabit this one more conservingly, we will see our lives, as well as the land, swiftly degraded. There is no contradiction between caring for our fellow human beings and caring for the rest of nature; on the contrary, only by attending to the health of the land can we measure our true needs or secure a lasting home.

Just before Eva was to leave again for college, she and I went for a hike in a nature preserve along Clear Creek, near Bloomington, to look at the hepatica and bloodroot and listen to the spring-high water. At the edge of the preserve, a wooden sign declared that the riding of horses was prohibited. The trail had been freshly gouged by horseshoes and was eroding badly. Trash snagged in the roots of sycamores along the stream. Much of the soil and its cover of wild-flowers had washed from the slopes where people scrambled up to picnic on the limestone bluff. Some of the cans they had left behind glinted among white stars of trillium.

I wondered what it would take to persuade the riders to get down off their horses and go on foot. What would it take to discourage people from dumping their worn-out washing machines in ditches? What would convince farmers to quit spraying poisons on their fields, suburbanites to quit spraying poisons on their lawns? What power in heaven or earth could stop loggers from seeing every tree as lumber, stop developers from seeing every acre of land as real estate, stop oil-company executives from seeing our last few scraps of wilderness as pay dirt waiting to be drilled? What would it take to persuade all of us to eat what we need, rather than what we can hold; to buy what we need, rather than what we can afford; to draw our pleasure from inexhaustible springs?

Signs will not work that change of mind, for in a battle of signs the billboards always win. Police cannot enforce it. Tongue lashings and sermons and earnest essays will not do it, nor will laws alone bring it about. The framers of the Constitution may have assumed that we did not need a Bill of Responsibilities because religion and reason and the benign impulses of our nature would lead us to care for one another and for our home. Having concluded a bloody century, and at the dawn of a new millennium that threatens to be still bloodier, few of us now feel much confidence in those redeeming influences. Only a powerful ethic might restrain us, retrain us, restore us. Our survival is at stake, yet worrying about our survival alone is too selfish a motive to carry us as far as we need to go. Nothing short of reimagining where we are and what we are will be radical enough to transform how we live.

Aldo Leopold gave us the beginnings of this new ethic half a century ago, in *A Sand County Almanac,* where he described the land itself as a community made up of rock, water, soil, plants, and animals—including *Homo sapiens,* the only species clever enough to ignore, for a short while, the conditions of membership. "We abuse land because we see it as a commodity belonging to us," Leopold wrote. "When we see land as a community to which we belong, we may begin to use it with love and respect." To use our places with love and respect demands from us the same generosity and restraint that we show in our dealings with a wife or husband, a child or parent, a neighbor, a stranger in trouble.

Once again this spring, the seventy-seventh of her life, my mother put out lint from her clothes dryer for the birds to use in building their nests. "I know how hard it is to make a home from scratch," she says, "I've done it often enough myself." That is not anthropomorphism; it is fellow feeling, the root of all kindness.

Doctors the world over study the same physiology, for we are one species, woven together by strands of DNA that stretch back to the beginnings of life. There is, in fact, only one life, one pulse animating the dust. Sycamores and snakes, grasshoppers and grass, hawks and humans all spring from the same source and all return to it. We need to make of this common life not merely a metaphor, although we live by metaphors, and not merely a story, although we live by stories; we need to make the common life a fact of the heart. That awareness, that concern, that love needs to go as deep in us as the feeling we have when a child dashes into the street and we hear a squeal of brakes, or when a piece of our home ground goes under concrete, or when a cat purrs against our palms or rain sends shivers through our bones or a smile floats back to us from another face.

With our own population booming, with frogs singing in the April ponds, with mushrooms cracking the pavement, life may seem the most ordinary of forces, like hunger or gravity; it may seem cheap, since we see it wasted every day. But in truth life is expensive, life is extraordinary, having required five billion years of struggle and luck on one stony, watery planet to reach its present precarious state. And so far as we have been able to discover by peering out into the great black spaces, the life that is common to all creatures here on earth is exceedingly uncommon, perhaps unique, in the universe. How, grasping that, can we remain unchanged?

It may be that we will not change, that nothing can restrain us, that we are incapable of reimagining our relations to one another or our place in creation. So many alarm bells are ringing, we may be tempted to stuff our ears with cotton. Better that we should keep ears and eyes open, take courage as well as joy from our common life, and work for what we love. What I love is curled about a loaf of bread, a family, a musical neighbor, a building salvaged for art, a town of familiar faces, a creek and a limestone bluff and a sky full of birds. Those may seem like frail threads to hold anyone in place while history heaves us about, and yet, when they are braided together, I find them to be amazingly strong.

Chapter 15

The Boundary

Wendell Berry

Kentucky farmer Wendell Berry, introduced earlier with his essay "The Whole Horse," is the author of more than three dozen books, including major works of poetry, fiction, and nonfiction. Although all his writings are much admired, many readers find his ideas and sentiments expressed most vividly in his several novels and short story collections—all set in a fictionalized version of his local Kentucky landscape, on the farms, forests, hills, and draws around his hometown. As visitors to this richly imagined world, we come to know well the several interconnected farm families that reside here, the bad as well as the good, the adherents of agrarian values as well as those gripped by wanderlust and the market's lure. Although the setting is contained, Port William harbors all the forces of the modern age, particularly those that have pressed so hard against the land and its human inhabitants.

One recurring character in Berry's fiction is Mat Feltner, a venerable tender of Port William's agrarian legacy. Berry admires Mat as much as he does any of his creations, and in him we see lived out the possibilities and strains of the agrarian way. The short story included here, taken from Berry's collection The Wild Birds, *is set in 1965, the final year of Mat's life. Now eighty, Mat finds himself with more acquaintances among the dead than among the living. His only son, Virgil—heir to his agrarian tradition—disappeared two decades earlier in the outburst of industrial destruction known as World War II. Although Mat has never recovered fully from the loss, Virgil's widow, Hannah, has remarried, and her new husband, Nathan Coulter, has become to Mat like a son. In this episode, Mat walks the boundaries of his longtime family farm. As*

*he does so, reflecting on the many lives linked to the place, he displays
with remarkable clarity how deeply and happily rooted a true agrarian
can be, on or off the farm.*

He can hear Margaret at work in the kitchen. That she knows well
what she is doing and takes comfort in it, one might tell from the
sounds alone as her measured, quiet steps move about the room. It
is all again as it has been during the almost twenty years that only
the two of them have lived in the old house. Sitting in the split-
hickory rocking chair on the back porch, Mat listens; he watches
the smoke from his pipe drift up and out past the foliage of the
plants in their hanging pots. He has finished his morning stint in
the garden, and brought in a half-bushel of peas that he set down
on the drainboard of the sink, telling Margaret, "There you are,
ma'am." He heard with pleasure her approval, "Oh! They're nice!"
and then he came out onto the shady porch to rest.

Since winter he has not felt well. Through the spring, while
Nathan and Elton and the others went about the work of the fields,
Mat, for the first time, confined himself to the house and barn lot
and yard and garden, working a little and resting a little, finding it
easier than he expected to leave the worry of the rest of it to
Nathan. But slowed down as he is, he has managed to make a difference. He has made the barn his business, and it is cleaner and in
better order than it has been for years. And the garden, so far, is
nearly perfect, the best he can remember. By now, in the first week
of June, in all its green rows abundance is straining against order.
There is not a weed in it. Though he has worked every day, he has
had to measure the work out in little stints, and between stints he
has had to rest.

But rest, this morning, has not come to him. When he went out
after breakfast he saw Nathan turning the cows and calves into the
Shade Field, so called for the woods that grows there on the slope
above the stream called Shade Branch. He did not worry about it
then, or while he worked through his morning jobs. But when he
came out onto the porch and sat down and lit his pipe, a thought

that had been on its way toward him for several hours finally reached him. He does not know how good the line fence is down Shade Branch; he would bet that Nathan, who is still rushing to get his crops out, has not looked at it. The panic of a realized neglect came upon him. It has been years since he has walked that fence himself, and he can see in his mind, as clearly as if he were there, perhaps five places where the winter spates of Shade Branch might have torn out the wire.

He sits, listening to Margaret, looking at pipe smoke, anxiously working his way down along that boundary in his mind.

"Mat," Margaret says at the screen door, "dinner's ready."

"All right," he says, though for perhaps a minute after that he does not move. And then he gets up, steps to the edge of the porch to knock out his pipe, and goes in.

❊   ❊   ❊

When he has eaten, seeing him pick up his hat again from the chair by the door, Margaret says, "You're not going to take your nap?"

"No," he says, for he has decided to walk that length of the boundary line that runs down Shade Branch. And he has stepped beyond the feeling that he is going to do it because he should. He is going to do it because he wants to. "I got something yet I have to do."

He means to go on out the door without looking back. But he knows that she is watching him, worried about him, and he goes back to her and gives her a hug. "It's all right, my old girl," he says. He stands with his arms around her, who seems to him to have changed almost while he has held her from girl to wife to woman to mother to grandmother to great-grandmother. There in the old room where they have been together so long, ready again to leave it, he thinks, "I am an old man now."

"Don't worry," he says. "I'm feeling good."

He does feel good, for an old man, and once outside, he puts the house behind him and his journey ahead of him. At the barn he takes from its nail in the old harness room a stout stockman's cane.

He does not need a cane yet, and he is proud of it, but as a concession to Margaret he has decided to carry one today.

When he lets himself out through the lot gate and into the open, past the barn and the other buildings, he can see the country lying under the sun. Nearby, on his own ridges, the crops are young and growing, the pastures are lush, a field of hay has been raked into curving windrows. Inlets of woods, in the perfect foliage of the early season, reach up the hollows between the ridges. Lower down, these various inlets join in the larger woods embayed in the little valley of Shade Branch. Beyond the ridges and hollows of the farm he can see the opening of the river valley, and beyond that the hills on the far side, blue in the distance.

He has it all before him, this place that has been his life, and how lightly and happily now he walks out again into it! It seems to him that he has cast off all restraint, left all encumbrances behind, taking only himself and his direction. He is feeling good. There has been plenty of rain, and the year is full of promise. The country looks promising. He thinks of the men he knows who are at work in it: the Coulter brothers and Nathan, Nathan's boy, Mattie, Elton Penn, and Mat's grandson, Bess's and Wheeler's boy, Andy Catlett. They are at Elton's now, he thinks, but by midafternoon they should be back here, baling the hay.

Carrying the cane over his shoulder, he crosses two fields, and then, letting himself through a third gate, turns right along the fencerow that will lead him down to Shade Branch. Soon he is walking steeply downward among the trunks of trees, and the shifting green sea of their foliage has closed over him.

❧   ❧   ❧

He comes into the deeper shade of the older part of the woods where there is little browse and the cattle seldom come, and here he sits down at the root of an old white oak to rest. As many times before, he feels coming to him the freedom of the woods, where he has no work to do. He feels coming to him such rest as, bound to house and barn and garden for so long, he had forgot. In body, now, he is an old man, but mind and eye look out of his old body into the shifting leafy lights and shadows among the still trunks with a recognition that is without age, the return of an ageless joy. He needs the rest, for he has walked in his gladness at a faster pace than he is used to, and he is sweating. But he is in no hurry, and he sits and grows quiet among the sights and sounds of the place. The time of the most abundant blooming of the woods flowers is past now, but the tent villages of mayapple are still perfect, there are ferns and stonecrop, and near him he can see the candle-like white flowers of black cohosh. Below him, but still out of sight, he can hear the water in Shade Branch passing down over the rocks in a hundred little rapids and falls. When he feels the sweat beginning to dry on his face he gets up, braces himself against the gray trunk of the oak until he is steady, and stands free. The descent beckons and he yields eagerly to it, going on down into the tireless chanting of the stream.

He reaches the edge of the stream at a point where the boundary, coming down the slope facing him, turns at a right angle and follows Shade Branch in its fall toward the creek known as Willow Run. Here the fence that Mat has been following crosses the branch over the top of a rock wall that was built in the notch of the stream long before Mat was born. The water coming down, slowed by the wall, has filled the notch above it with rock and silt, and then, in freshet, leaping over it, has scooped out a shallow pool below it, where water stands most of the year. All this, given the continuous little changes of growth and wear in the woods and the stream, is as it was when Mat first knew it: the wall gray and mossy, the water, only a spout now, pouring over the wall into the little pool, covering the face of it with concentric wrinkles sliding outward.

Here, seventy-five years ago, Mat came with a fencing crew: his

father, Ben, his uncle, Jack Beechum, Joe Banion, a boy then, not
much older than Mat, and Joe's grandfather, Smoke, who had been
a slave. And Mat remembers Jack Beechum coming down through
the woods, as Mat himself has just come, carrying on his shoulder
two of the long light rams they used to tamp the dirt into postholes.
As he approached the pool he took a ram in each hand, holding
them high, made three long approaching strides, planted the rams
in the middle of the pool, and vaulted over. Mat, delighted, said,
"Do it again!" And without breaking rhythm, Jack turned, made
the three swinging strides, and did it again—*does* it again in Mat's
memory, so clearly that Mat's presence there, so long after, fades
away, and he hears their old laughter, and hears Joe Banion say,
"Mistah Jack, he might nigh a *bird*!"

Forty-some years later, coming down the same way to build that
same fence again, Mat and Joe Banion and Virgil, Mat's son, grown
then and full of the newness of his man's strength, Mat remembered what Jack had done and told Virgil; Virgil took the two rams,
made the same three strides that Jack had made, vaulted the pool,
and turned back and grinned. Mat and Joe Banion laughed again,
and this time Joe looked at Mat and said only "Damn."

Now a voice in Mat's mind that he did not want to hear says,
"Gone. All of them are gone." And they *are* gone. Mat is standing
by the pool, and all the others are gone, and all that time has passed.
And still the stream pours into the pool and the circles slide across
its face.

❃   ❃   ❃

He shrugs as a man would shake snow from his shoulders and steps
away. He finds a good place to cross the branch, and picks his way
carefully from rock to rock to the other side, using the cane for that
and glad he brought it. Now he gives attention to the fence. Soon
he comes upon signs—new wire spliced into the old, a staple newly

driven into a sycamore—that tells him his fears were unfounded. Nathan has been here. For a while now Mat walks in the way he knows that Nathan went. Nathan is forty-one this year, a quiet, careful man, as attentive to Mat as Virgil might have been if Virgil had lived to return from the war. Usually, when Nathan has done such a piece of work as this, he will tell Mat so that Mat can have the satisfaction of knowing that the job is done. Sometimes, though, when he is hurried, he forgets, and Mat will think of the job and worry about it and finally go to see to it himself, almost always to find, as now, that Nathan has been there ahead of him and has done what needed to be done. Mat praises Nathan in his mind and calls him son. He has never called Nathan son aloud, to his face, for he does not wish to impose or intrude. But Nathan, who is not his son, has become his son, just as Hannah, Nathan's wife, Virgil's widow, who is not Mat's daughter, has become his daughter.

"I am blessed," he thinks. He walks in the way Nathan walked down along the fence, between the fence and the stream, seeing Nathan in his mind as clearly as if he were following close behind him, watching. He can see Nathan with axe and hammer and pliers and pail of staples and wire stretcher and coil of wire, making his way down along the fence, stopping now to chop a windfall off the wire and retighten it, stopping again to staple the fence to a young sycamore that has grown up in the line opportunely to serve as a post. Mat can imagine every move Nathan made, and in his old body, a little tired now, needing to be coaxed and instructed in the passing of obstacles, he remembers the strength of the body of a man of forty-one, unregarding of its own effort.

Now, trusting the fence to Nathan, Mat's mind turns away from it. He allows himself to drift down the course of the stream, passing through it as the water passes, drawn by gravity, bemused by its little chutes and falls. He stops beside one tiny quiet backwater and watches a family of water striders conducting their daily business, their feet dimpling the surface. He eases the end of his cane into the pool, and makes a crawfish spurt suddenly backward beneath a rock.

A water thrush moves down along the rocks of the streambed ahead of him, teetering and singing. He stops and stands to watch while a large striped woodpecker works its way up the trunk of a big sycamore, putting its eye close to peer under the loose scales of the bark. And then the bird flies to its nesting hole in a hollow snag still nearer by to feed its young, paying Mat no mind. He has become still as a tree, and now a hawk suddenly stands on a limb close over his head. The hawk loosens his feathers and shrugs, looking around him with his fierce eyes. And it comes to Mat that once more, by stillness, he has passed across into the wild inward presence of the place.

"Wonders," he thinks. "Little wonders of a great wonder." He feels the sweetness of time. If a man eighty years old has not seen enough, then nobody will ever see enough. Such a little piece of the world as he has before him now would be worth a man's long life, watching and listening. And then he could go two hundred feet and live again another life, listening and watching, and his eyes would never be satisfied with seeing, nor his ears filled with hearing. Whatever he saw could be seen only by looking away from something else equally worth seeing. For a second he feels and then loses some urging of the delight in a mind that could see and comprehend it all, all at once. "I could stay here a long time," he thinks. "I could stay here a long time."

❊   ❊   ❊

He is standing at the head of a larger pool, another made by the plunging of the water over a rock wall. This one he built himself, he and Virgil, in the terribly dry summer of 1930. By the latter part of that summer, because of the shortage of both rain and money, they had little enough to do, and they had water on their minds.

Mat remembered this place, where a strong vein of water opened under the roots of a huge old sycamore and flowed only a few feet before it sank uselessly among the dry stones of the streambed. "We'll make a pool," he said. He and Virgil worked several days in the August of that year, building the wall and filling in behind it so that the stream, when it ran full again, would not tear out the stones. The work there in the depth of the woods took their minds off their parched fields and comforted them. It was a kind of work that Mat loved and taught Virgil to love, requiring only the simplest tools: a large sledgehammer, a small one, and two heavy crowbars with which they moved the big, thick rocks that were in that place. Once their tools were there, they left them until the job was done. When they came down to work they brought only a jug of water from the cistern at the barn.

"We could drink out of the spring," Virgil said.

"Of course we could," Mat said. "It's dog days now. Do you want to get sick?"

In a shady place near the creek, Virgil tilted a flagstone up against a small sycamore, wedging it between trunk and root, to make a place for the water jug. There was not much reason for that. It was a thing a boy would do, making a little domestic nook like that, so far off in the woods, but Mat shared his pleasure in it, and that was where they kept the jug.

When they finished the work and carried their tools away, they left the jug, forgot it, and did not go back to get it. Mat did not think of it again until, years later, he happened to notice the rock still leaning against the tree, which had grown over it, top and bottom, fastening the rock to itself by a kind of natural mortise. Looking under the rock, Mat found the earthen jug still there, though it had been broken by the force of the tree trunk growing against it. He left it as it was. By then Virgil was dead, and the stream, rushing over the wall they had made, had scooped out a sizable pool that had been a faithful water source in dry years.

Remembering, Mat goes to the place and looks and again finds the stone and finds the broken jug beneath it. He has never touched rock or jug, and he does not do so now. He stands, looking, thinking of his son, dead twenty years, a stranger to his daughter, now a grown woman, who never saw him, and he says aloud, "Poor fellow!" So taken is his mind by his thoughts that he does not know he is weeping until he feels his tears cool on his face.

Deliberately, he turns away. Deliberately, he gives his mind back to the day and the stream. He goes on down beside the flowing water, loitering, listening to the changes in its voice as he walks along it. He silences his mind now and lets the stream speak for him, going on, descending with it, only to prolong his deep peaceable attention to that voice that speaks always only of where it is, remembering nothing, fearing and desiring nothing.

Farther down, the woods thinning somewhat, he can see ahead of him where the Shade Branch hollow opens into the valley of Willow Run. He can see the crest of the wooded slope on the far side of the creek valley. He stops. For a minute or so his mind continues on beyond him, charmed by the juncture he has come to. He imagines the succession of them, openings on openings: Willow Run opening to the Kentucky River, the Kentucky to the Ohio, the Ohio to the Mississippi, the Mississippi to the Gulf of Mexico, the Gulf to the boundless sky. He walks in old memory out into the river, carrying a heavy rock in each hand, out and down, until the water closes over his head and then the light shudders above him and disappears, and he walks in the dark cold water, down the slant of the bottom, to the limit of breath, and then drops his weights and cleaves upward into light and air again.

❧   ❧   ❧

He turns around and faces the way he has come. "*Well,* old man!" he thinks. "*Now* what are you going to do?" For he has come down a long way, and now, looking back, he feels the whole country tilted against him. He feels the weight of it and the hot light over it. He hears himself say aloud, "Why, I've got to get back out of here."

But he is tired. It has been a year since he has walked even so far as he has already come. He feels the heaviness of his body, a burden that, his hand tells him, he has begun to try to shift off his legs and onto the cane. He thinks of Margaret, who, he knows as well as he knows anything, already wonders where he is and is worrying about him. Fear and exasperation hold him for a moment, but he pushes them off; he forces himself to be patient with himself. "Well," he says, as if joking with Virgil, for Virgil has come back into his thoughts now as a small boy, "going up ain't the same as coming down. It's going to be different."

It would be possible to go on down, and he considers that. He could follow the branch on down to the Willow Run road where it passes the Rowanberry place. That would be downhill, and if he could find Mart Rowanberry near the house, Mart would give him a ride home. But the creek road is little traveled these days; if he goes down there and Mart is up on the ridge at work, which he probably is, then Mat will be farther from home than he is now, and will have a long walk, at least as far as the blacktop, maybe farther. Of course, he could go down there and just wait until Mart comes to the house at quitting time. There is sense in that, and for a moment Mat stands balanced between ways. What finally decides him is that he is unsure what lies between him and the creek road. If he goes down much farther he will cross the line fence onto the Rowanberry place. He knows that he would be all right for a while, going on down along the branch, but once in the creek bottom he would have to make his way to the road through dense, under-growthy thicket, made worse maybe by piles of drift left by the winter's high water. He might have trouble getting through there, he thinks, and the strangeness of that place seems to forbid him. It has begun to trouble him that no other soul on earth knows where he is. He does not want to go where he will not know where he is himself.

❧   ❧   ❧

He chooses the difficult familiar way, and steps back into it, helping himself with the cane now. He does immediately feel the difference between coming down and going up, and he wanders this way and that across the line of his direction, searching for the easiest steps. Windfalls that he went around or stepped over thoughtlessly, coming down, now require him to stop and study and choose. He is tired. He moves by choice.

He and his father have come down the branch, looking for a heifer due to calve; they have found her and are going back. Mat is tired. He wants to be home, but he does not want to *go* home. He is hot and a scratch on his face stings with sweat. He would just as soon cry as not, and his father, walking way up ahead of him, has not even slowed down. Mat cries, "*Wait,* Papa!" And his father does turn and wait, a man taller than he looks because of the breadth of his shoulders, whom Mat would never see in a hurry and rarely see at rest. He has turned, smiling in the heavy bush of his beard, looking much as he will always look in Mat's memory of him, for Mat was born too late to know him young and he would be dead before he was old. "Come on, Mat," Ben says. "Come on, my boy." As Mat comes up to him, he reaches down with a big hand that Mat puts his hand into. "It's all right. It ain't that far." They go on up the branch then. When they come to a windfall across the branch, Ben says, "This one we go under." And when they come to another, he says, "This one we go around."

Mat, who came down late in the afternoon to fix the fence, has fixed it, and is hurrying back, past chore time, and he can hear Virgil behind him, calling to him, "Wait, Daddy!" He brought Virgil against his better judgment, because Virgil would not be persuaded

not to come. "You need me," he said. "I do need you," Mat said,
won over. "You're my right-hand man. Come on." But now, irri-
tated with himself and with Virgil too, he knows that Virgil needs
to be carried, but his hands are loaded with tools and he *can't* carry
him. Or so he tells himself, and he walks on. He stretches the distance out between them until Virgil feels that he has been left alone
in the darkening woods; he sits down on a rock and gives himself
up to grief. Hearing him cry, Mat puts his tools down where he can
find them in the morning, and goes back for Virgil. "Well, it's all
right, old boy," he says, picking him up. "It's all right. It's all right."

He is all right, but he is sitting down on a tree trunk lying across
the branch, and he has not been able to persuade himself to get up.
He came up to the fallen tree, and, to his surprise, instead of step-
ping over it, he sat down on it. At first that seemed to him the
proper thing to do. He needed to sit down. He was tired. But now
a protest begins in his mind. He needs to be on his way. He ought
to be home by now. He knows that Margaret has been listening for
him. He knows that several times by now she has paused in her
work and listened, and gone to the windows to look out. She is
hulling the peas he brought her before dinner. If he were there, he
would be helping her. He thinks of the two of them sitting in the
kitchen, hulling peas, and talking. Such a sense of luxury has come
into their talk, now that they are old and in no hurry. They talk of
what they know in common and do not need to talk about, and so
talk about only for pleasure.

They would talk about where everybody is today and what each
one is doing. They would talk about the stock and the crops. They
would talk about how nice the peas are this year, and how good the
garden is.

He thought, once, that maybe they would not have a garden.
There were reasons not to have one.

"We don't need a garden this year," he said to Margaret, wanting to spare her the work that would be in it for her.

"Yes," she said, wanting to spare him the loss of the garden, "of course we do!"

"Margaret, we'll go to all that work, and can all that food, and neither one of us may live to eat it."

She gave him her smile, then, the same smile she had always given him, that always seemed to him to have survived already the worst he could think of. She said, "Somebody will."

She pleased him, and the garden pleased him. After even so many years, he still needed to be bringing something to her.

※　※　※

His command to get up seems to prop itself against his body and stay there like a brace until finally, in its own good time and again to his surprise, his body obeys. He gets up, steps over the tree, and goes on. He keeps himself on his feet for some time now, herding himself along like a recalcitrant animal, searching for the easy steps, reconciling himself to the hard steps when there are no easy ones. He is sweating heavily. The air is hot and close, so deep in that cleft of the hill. He feels that he must stretch upward to breathe. It is as though his body has come to belong in a different element, and the mere air of that place now hardly sustains it.

He comes to the pool, the wall that he and Virgil made, and pauses there. "Now you must drink," he says. He goes to where the spring comes up among the roots of the sycamore. There is a smooth clear pool there, no bigger than his hat. He lies down to drink, and drinks, looking down into the tiny pool cupped among the roots, surrounded with stonecrop and moss. The loveliness of it holds him: the cool water in that pretty place in the shade, the great tree rising and spreading its white limbs overhead. "I am blessed," he thinks, "I could stay here." He rests where he lies, turned away from his drinking, more comfortable on the roots and rocks than he expected to be. Through the foliage he can see white clouds mov-

ing along as if mindful where they are going. A chipmunk comes
in quick starts and stops across the rocks and crouches a long time
not far from Mat's face, watching him, as if perhaps it would like a
drink from the little pool that Mat just drank from. "Come on,"
Mat says to it. "There ain't any harm in me." He would like to
sleep. There is a weariness beyond weariness in him that sleep
would answer. He can remember a time when he could have let
himself sleep in such a place, but he cannot do that now. "Get up,"
he says aloud. "Get up, get up."

But for a while yet he does not move. He and the chipmunk
watch each other. Now and again their minds seem to wander
apart, and then they look again and find each other still there.
There is a sound of wings, like a sudden dash of rain, and the chipmunk tumbles off its rock and does not appear again. Mat laughs.
"You'd *better* hide." And now he does get up.

He stands, his left hand propped against the trunk of the
sycamore. Darkness draws across his vision and he sinks back
down onto his knees, his right hand finding purchase in the old cup
of the spring. The darkness wraps closely around him for a time,
and then withdraws, and he stands again. "*That* won't do," he says
to Virgil. "We got to do better than that, boy." And then he sees his
father too standing with Virgil on the other side of the stream.
They recognize him, even though he is so much older now than
when they knew him—older than either of them ever lived to be.
"Well," he says, "looks like we got plenty of help." He reaches
down and lets his right hand feel its way to the cane, picks it up, and
straightens again. "*Yes*sir."

The world clears, steadies, and levels itself again in the light. He
looks around him at the place: the wall, the pool, the spring mossy
and clear in the roots of the white tree. "I am not going to come
back here," he thinks. "I will never be in this place again."

Instructing his steps, he leaves. He moves with the utmost care
and the utmost patience. For some time he does not think of where

he is going. He is merely going up along the stream, asking first for one step and then for the next, moving by little plans that he carefully makes, and by choice. When he pauses to catch his breath or consider his way, he can feel his heart beating; at each of its beats the world seems to dilate and spring away from him.

His father and Virgil are with him, moving along up the opposite side of the branch as he moves up his side. He cannot always see them, but he knows they are there. First he does not see them, and then he sees one or the other of them appear among the trees and stand looking at him. They do not speak, though now and again he speaks to them. And then Jack Beechum, Joe Banion, and old Smoke are with them. He sees them sometimes separately, sometimes together. The dead who were here with him before are here with him again. He is not afraid. "I could stay here," he thinks. But ahead of him there is a reason he should not do that, and he goes on.

He seems to be walking in and out of his mind. Or it is time, perhaps, that he is walking in and out of. Sometimes he is with the dead as they were, and he is as he was, and all of them together are walking upward through the woods toward home. Sometimes he is alone, an old man in a later time than any of the dead have known, going the one way that he alone is going, among all the ways he has gone before, among all the ways he has never gone and will never go.

❋   ❋   ❋

He does not remember falling. He is lying on the rocks beside the branch, and there is such disorder and discomfort in the way he is lying as he could not have intended. And so he must have fallen. He wonders if he is going to get up. After a while he does at least sit up. He shifts around so that his back can rest against the trunk of a tree. His movements cause little lurches in the world, and he waits for it to be steady. "Now you have got to stop and think," he says. And then he says, "Well, you have stopped. Now you had better think."

He does begin to think, forcing his vision and his thoughts out away from him into the place around him, his mind making little articulations of recognition. The place and his memory of it begin to speak to one another. He has come back almost to the upper wall and pool, where he first came down to the branch. When he gets there he will have a choice to make between two hard ways to go.

But his mind, having thought of those choices, now leaves him, like an undisciplined pup, and goes to the house, and goes back in time to the house the way it was when he and Margaret were still young, when Virgil was four or five years old, and Bess was eleven or twelve.

About this time in the afternoon, about this time in the year, having come to the house for something, he cannot remember what, he pushes open the kitchen door, leans his shoulder on the jamb, and looks at Margaret who stands with her back to him, icing a cake.

"Now nobody's asked for my opinion," he says, "and nobody's likely to, but if anybody ever was to, I'd say that *that* is a huggable woman."

"Don't you come near me, Mat Feltner."

"And a spirited woman."

"If you so much as lay a hand on me, I'm going to hit you with this cake."

"And a dangerous, mean woman."

"Go back to work."

"Who is, still and all, a huggable woman. Which is only my opinion. A smarter man might think different."

She turns around, laughing, and comes to get her hug. "I could never have married a smart man."

❉ ❉ ❉

"She didn't marry *too* smart a one," he thinks. He is getting up, the effort requiring the attendance of his mind, and once he is standing he puts his mind back on his problem. That is not where it wants to be, but this time he makes it stay. If he leaves the branch and goes back up onto the ridge by the way he came down, that will require a long slanting climb up across the face of a steep slope. "And it has got steeper since I came down it," he thinks. If he goes on up Shade Branch, which would be the easiest, surest route, he will have, somehow, to get over or through the fence that crosses the branch above the wall. He does not believe that he can climb the fence. Where the fence crosses the stream it is of barbed wire, and in that place a stronger man might go through or under it. But he does not want to risk hooking his clothes on the barbs.

But now he thinks of a third possibility: the ravine that comes into Shade Branch just above him to his right hand. The dry, rocky streambed in the ravine would go up more gently than the slope, the rocks would afford him stairsteps of a sort for at least some of the way, and it would be the shortest way out of the woods. It would bring him out farther from home than the other ways, but he must not let that bother him. It is the most possible of the three ways, and the most important thing now, he knows, is to get up onto the open land of the ridge where he can be seen if somebody comes looking for him. Somebody will be looking for him, he hopes, for he has to admit that he is not going very fast, and once he starts up the ravine he will be going slower than ever.

For a while he kept up the belief, and then the hope, that he could make it home in a reasonable time and walk into the house as if nothing had happened.

"Where on earth have you been?" Margaret would ask.

He would go to the sink to wash up, and then he would say, drying his hands, "Oh, I went to see about the line fence down the branch, but Nathan had already fixed it."

He would sit down then to help her with the peas.

"Mat Feltner," she would say, "surely you didn't go away off down there."

But it is too late now, for something *has* happened. He has been gone too long, and is going to be gone longer.

Margaret has got up from her work and gone to the windows and looked out, and gone to the door onto the back porch and spoken his name, and walked on out to the garden gate and then to the gate to the barn lot.

He can see her and hear her calling as plainly as if he were haunting her. "Mat! Oh, Mat!"

He can hear her, and he makes his way on up the branch to the mouth of the ravine. He turns up the bed of the smaller stream. The climb is steeper here, the hard steps closer together. The ascent asks him now really to climb, and in places, where the rocks of the streambed bulge outward in a wall, he must help himself with his hands. He must stoop under and climb over the trunks of fallen trees. When he stops after each of these efforts the heavy beating of his heart keeps on. He can feel it shaking him, and darkness throbs in his eyes. His breaths come too far between and too small. Sometimes he has roots along the side of the ravine for a banister, and that helps. Sometimes the cane helps; sometimes, when he needs both hands, it is in the way. And always he is in the company of the dead.

❊ ❊ ❊

Ahead of him the way is closed by the branchy top of a young maple, blown down in a storm, and he must climb up the side of the ravine to get around it. At the top of the climb, when the slope has gentled and he stops and his heart plunges on and his vision darkens, it seems to him that he is going to fall; he decides instead to sit down, and does. Slowly he steadies again within himself. His heart slows and his vision brightens again. He tells himself again to get up. "It ain't as far as it has been," he says to Virgil. "I'm going to be all right now. I'm going to make it."

But now his will presses against his body, as if caught within it, in bewilderment. It will not move. There was a time when his body had strength enough in it to carry him running up such a place as this, with breath left over to shout. There was a time when it had barely enough strength in it to carry it this far. There is a time when his body is too heavy for his strength. He longs to lie down. To Jack Beechum, the young man, Jack Beechum, who is watching him now, he says, "You and I were here once."

The dead come near him, and he is among them. They come and go, appear and disappear, like a flock of feeding birds. They are there and gone. He is among them, and then he is alone. To one who is not going any farther, it is a pretty place, the leaves new and perfect, a bird singing out of sight among them somewhere over his head, and the softening light slanting in long beams from the west. "I could stay here," he thinks. It is the thought of going on that turns that steep place into an agony. His own stillness pacifies it and makes it lovely. He thinks of dying, secretly, by himself, in the woods. No one now knows where he is. Perhaps it would be possible to hide and die, and never be found. It would be a clean, clear way for that business to be done, and the thought, in his weariness, comforts him, for he has feared that he might die a nuisance to Margaret and the others. He might, perhaps, hide himself in a little cave or sink hole if one were nearby, here where the dead already are, and be one of them, and enter directly into the peaceableness of this place, and turn with it through the seasons, his body grown easy in its weight.

❀　❀　❀

But there is no hiding place. He would be missed and hunted for and found. He would die a nuisance, for he could not hide from all the reasons that he would be missed and worried about and hunted for. He has an appointment that must be kept, and between him and it the climb rises on above him.

He has an accounting he must come to, and it is not with the dead, for Margaret has not sat down again, but is walking. She is walking from room to room and from window to window. She has not called Bess, because she does not want Bess to drive all the way up from Hargrave, perhaps for nothing. Though she has thought about it, she has not even called Hannah, who is nearer by. She does not want to alarm anybody. But she is alarmed. She walks from room to room and from window to window, pausing to look out, and walking again. She walks with her arms tightly folded, as she has walked all her life when she has been troubled, until Mat, watching her, has imagined that he thinks as she thinks and feels as she feels, so moved by her at times that he has been startled to realize again his separateness from her.

He remembers the smile of assent that she gave him once: "Why, Mat, I thought you did. And I love you." Everything that has happened to him since has come from that—and leads to that, for it is not a moment that has ever stopped happening; he has gone toward it and aspired to it all his life, a time that he has not surpassed.

Now she is an old woman, walking in his mind in the rooms of their house. She has called no one and told no one. She is the only one who knows that she does not know where he is. The men are in the hayfield, and she is waiting for one of them or some of them to come to the barn. Or Wheeler might come by. It is the time of day when he sometimes does. She walks slowly from room to room, her arms folded tightly, and she watches the windows.

Mat, sitting in his heaviness among the trees, she does not know where, yearns for her as from beyond the grave. "Don't worry," he says. "It's going to be all right." He gets up.

And now an overmastering prayer that he did not think to pray rushes upon him out of the air and seizes him and grapples him to itself: an absolute offering of himself to his return. It is an offer, involuntary as his breath, voluntary as the new steps he has already taken up the hill, to give up his life in order to have it. The prayer does not move him beyond weariness and weakness, it moves him merely beyond all other thoughts.

❊   ❊   ❊

He gives no more regard to death or to the dead. The dead do not appear again. Now he is walking in this world, walking in time, going home. A shadowless love moves him now, not his, but a love that he belongs to, as he belongs to the place and to the light over it. He is thinking of Margaret and of all that his plighting with her has led to. He is thinking of the membership of the fields that he has belonged to all his life, and will belong to while he breathes, and afterward. He is thinking of the living ones of that membership— at work today in the fields that the dead were at work in before them.

"I am blessed," he thinks. "I am blessed."

He is crawling now, the cane lost and forgotten. He crawls a little, and he rests a lot. The slope has gentled somewhat. The big woods has given way to thicket. He has turned away from the stream, taking the straightest way to the open slope that he can see not far above him. The cattle are up there grazing, the calves starting to play a little, now that the cool of the day is here.

When he comes out, clear of the trees, onto the grassed hillside, he seems again to have used up all his strength. "Now," he thinks, "you have got to rest." Once he has rested he will go on to the top of the ridge. Once he gets there, he can make it to the road. He crawls on up the slope a few feet to where a large walnut tree stands alone outside the woods, and sits against it so that he will have a prop for his back. He wipes his face, brushes at the dirt and litter on his knees and then subsides. Not meaning to, he sleeps.

❀   ❀   ❀

The sun is going down when he wakes, the air cold on his damp clothes. Except for opening his eyes, he does not move. His body is still as a stone.

Now he knows what woke him. It is the murmur of an automobile engine up on the ridge—Wheeler's automobile, by the sound of it. And when it comes into sight he sees that it is Wheeler's; Wheeler is driving and Elton Penn and Nathan are with him. They are not looking for him. They have not seen Margaret. Perhaps they did not bale the hay. Or they may have finished and got away early. But he knows that Wheeler found Nathan and Elton, wherever they were, after he shut his office and drove up from Hargrave, and they have been driving from field to field ever since, at Elton's place or at Nathan's or at Wheeler's, and now here. This is something they do, Mat knows, for he is often with them when they do it. Wheeler drives the car slowly, and they look and worry and admire and remember and plan. They have come to look at the cattle now, to see them on the new grass. They move among the cows and calves, looking and stopping. Now and then an arm reaches out of one of the car windows and points. For a long time they do not turn toward Mat. It is as though he is only part of the field, like the tree he is leaning against. He feels the strangeness of his stillness, but he does not move.

And then, still a considerable distance away, Wheeler turns the car straight toward the tree where Mat is sitting. He sees their heads go up when they see him, and he raises his right hand and gives them what, for all his eagerness, is only an ordinary little wave.

Wheeler accelerates, and the car comes tilting and rocking

down the slope. Where the slope steepens, forcing the car to slow down, Mat sees Nathan leap out of it and come running toward him, Elton out too, then, coming behind him, while Wheeler is still maneuvering the car down the hill. Seeing that they are running to help him, Mat despises his stillness. He forces himself to his knees and then to his feet. He turns to face Nathan, who has almost reached him. He lets go of the tree and stands, and sees the ground rising against him like a blow. He feels himself caught strongly, steadied, and held. He hears himself say, "Papa?"

❊　❊　❊

That night, when Margaret finds him wandering in the darkened house, he does not know where he is.

Notes

Introduction: A Durable Scale

1. A useful summary is James A. Montmarquet, *The Idea of Agrarianism: From Hunter-Gatherer to Agrarian Radical in Western Culture* (Moscow: University of Idaho Press, 1989).

2. See, e.g., Victor Davis Hanson, *The Land Was Everything: Letters from an American Farmer* (New York: Free Press, 2000); Wayne C. Rohrer and Louis H. Douglas, *The Agrarian Transition in America: Dualism and Change* (Indianapolis: Bobbs-Merrill, 1969); James Chen, "The American Ideology," *Vanderbilt Law Review* 48, no. 4 (1995): 809–77. On the various uses of the term, see Thomas P. Govan, "Agrarian and Agrarianism: A Study in the Use and Abuse of Words," *Journal of Southern History* 30 (1964): 35–47.

3. See Montmarquet, *Idea of Agrarianism,* 45–53.

4. Sir Albert Howard, *The Soil and Health: A Study of Organic Agriculture* (New York: Devin-Adair, 1947), 11.

5. Aldo Leopold, *A Sand County Almanac, and Sketches Here and There* (New York: Oxford University Press, 1949), 224–25. Leopold used the term *stability* as a synonym for the concept of land health, which he explored in various writings. See Eric T. Freyfogle, *"A Sand County Almanac* at Fifty: Leopold in the New Century," *Environmental Law Reporter* 30, no. 1 (January 2000): 10058–68.

6. Wendell Berry, *Another Turn of the Crank* (Washington, D.C.: Counterpoint Press, 1995), 90; Wendell Berry, *Sex, Economy, Freedom, and Community* (New York: Pantheon Books, 1993), 14–15, 40; Wendell

Berry, *What Are People For?* (San Francisco: North Point Press, 1990), 149, 206–7; Wendell Berry, *A Continuous Harmony: Essays on Culture and Agriculture* (New York: Harcourt Brace Jovanovich, 1972), 84, 164.

7. Allan Carlson, *The New Agrarian Mind: The Movement Toward Decentralist Thought in Twentieth-Century America* (New Brunswick, N.J.: Transaction, 2000), 88, 173, 196 ("nature worship").

8. Paul B. Thompson, *The Spirit of the Soil: Agriculture and Environmental Ethics* (London: Routledge, 1995), 3.

9. One vision of harmony is offered in Julia Freedgood, "Farming to Improve Environmental Quality," in *Visions of American Agriculture,* edited by William Lockeretz (Ames: Iowa State University Press, 1997), 77–90. A classic presentation is Aldo Leopold, "The Farmer as a Conservationist," *American Forests* 45 (June 1939): 6, reprinted in Aldo Leopold, *For the Health of the Land: Previously Unpublished Essays and Other Writings,* edited by J. Baird Callicott and Eric T. Freyfogle (Washington, D.C.: Island Press, Shearwater Books, 1999), 161–75.

10. Richard Weaver, "Two Types of American Individualism," in *The Southern Essays of Richard M. Weaver,* edited by George M. Curtis III and James J. Thompson Jr. (Indianapolis: Liberty Press, 1987), 77–103; Aldo Leopold, "The Farm Wildlife Program: A Self-Scrutiny" (manuscript, ca. 1937), Aldo Leopold Papers, University of Wisconsin–Madison Archives.

11. See, e.g., Wendell Berry, "The Work of Local Culture," in *What Are People For?* 153–69; Wendell Berry, "Does Community Have a Value?" in *Home Economics* (San Francisco: North Point Press, 1987), 179–92.

12. I paraphrase here Wendell Berry, "Preserving Wildness," in *Home Economics,* 146.

13. Wendell Berry, "Nature as Measure," in *What Are People For?* 204–10. Later, Wes Jackson used the phrase in "Nature as Measure," chap. 4 in *Becoming Native to This Place* (Lexington: University Press of Kentucky, 1994).

14. Wendell Berry has been perhaps most pointed in his criticisms; see Wendell Berry, "A Bad Big Idea," in *Sex, Economy, Freedom, and Community*, 45–51, and Wendell Berry, "The Whole Horse," in this volume.

15. See Wendell Berry, "Whose Head Is the Farmer Using? Whose Head Is Using the Farmer?" in *Meeting the Expectations of the Land: Essays in Sustainable Agriculture,* edited by Wes Jackson et al. (San Francisco: North Point Press, 1984), 19–30; Eric T. Freyfogle, "Ethics, Community, and Private Land," *Ecology Law Quarterly* 23 (1996): 631–61.

16. Evidence of the problem in terms of soil erosion is offered in William Harbaugh, "Twentieth-Century Tenancy and Soil Conservation: Some Comparisons and Questions," *Agricultural History* 66, no. 2 (spring 1992): 95–119.

17. Wendell Berry, "Private Property and the Common Wealth," in *Another Turn of the Crank,* 50–51; Richard Weaver, *Ideas Have Consequences* (Chicago: University of Chicago Press, 1948), 131–33.

18. Tim Lehman, *Public Values, Private Lands: Farmland Preservation Policy, 1933–1985* (Chapel Hill: University of North Carolina Press, 1995); Neal C. Gross, "A Post Mortem on County Planning," *Journal of Farm Economics* 25, no. 3 (1943): 644–61.

19. See Kathleen M. Merrigan, "Government Pathways to True Food Security," in Lockeretz, *Visions of American Agriculture,* 155–72.

20. Amitai Etzioni, *The New Golden Rule: Community and Morality in a Democratic Culture* (New York: Basic Books, 1996), 143.

21. A prominent manifesto of the Great Depression era was Frank Owsley, "The Pillars of Agrarianism," *American Review* 4 (March 1935): 529–47. See also Paul K. Conkin, *The Southern Agrarians* (Knoxville: University of Tennessee Press, 1988), 101–26.

22. See Robert Clark, ed., *Our Sustainable Table* (San Francisco: North Point Press, 1990); Kate Clancy, "Reconnecting Farmers and Citizens in the Food System," in Lockeretz, *Visions of American Agriculture,* 47–57.

23. Differing views are summarized in Leslie Stevenson, *Seven Theories of Human Nature,* 2d ed. (New York: Oxford University Press, 1987).

24. Etzioni, *New Golden Rule*, 165.

25. Berry, *Sex, Economy, Freedom, and Community*, 119–20.

26. Etzioni, *New Golden Rule*, 127.

27. Weaver, *Ideas Have Consequences*, 41.

28. Berry, *Sex, Economy, Freedom, and Community*, 172–73.

29. Hanson, *Land Was Everything*, 49–51.

30. See, e.g., Jane Adams, *The Transformation of Rural Life: Southern Illinois, 1890–1990* (Chapel Hill: University of North Carolina Press, 1994); Deborah Fink, *Agrarian Women: Wives and Mothers in Rural Nebraska, 1880–1940* (Chapel Hill: University of North Carolina Press, 1992); Wendell Berry, *The Hidden Wound* (San Francisco: North Point Press, 1989).

31. See Peggy Bartlett, "Farm Families in a Changing America," in Lockeretz, *Visions of American Agriculture*, 31–46.

32. Weaver, *Ideas Have Consequences*, 73.

33. See Joan Iverson Nassauer, "Agricultural Landscapes in Harmony with Nature," in Lockeretz, *Visions of American Agriculture*, 59–73.

34. Richard Weaver, "The South and the American Union," in Curtis and Thompson, *Southern Essays*, 239.

35. Louis Bromfield, *From My Experience: The Pleasures and Miseries of Life on a Farm* (New York: Harper & Brothers, 1955), 7.

36. A useful comparison of industrial farm values with communitarian ones is offered in Paul B. Thompson, "Agrarian Values: Their Future Place in U.S. Agriculture," in Lockeretz, *Visions of American Agriculture*, 17–30.

37. Montmarquet, *Idea of Agrarianism*, 87–88.

38. See Grant McConnell, *The Decline of Agrarian Democracy* (Berkeley: University of California Press, 1953).

39. See, e.g., Richard Hofstadter, *The Age of Reform* (New York: Random House, 1955), 43–46.

40. See David B. Danbom, "Past Visions of American Agriculture," in Lockeretz, *Visions of American Agriculture*, 3–16.

41. Daniel Bell, *The Cultural Contradictions of Capitalism* (New York:

Basic Books, 1976); Joseph A. Schumpeter, *Capitalism, Socialism, and Democracy* (New York: Harper & Brothers, 1942).

42. Eugene Genovese, *The Southern Tradition: The Achievement and Limitations of an American Conservatism* (Cambridge, Mass.: Harvard University Press, 1994), 38.

43. A recent survey from a conservation perspective is in Kirkpatrick Sale, *Rebels against the Future: The Luddites and Their War on the Industrial Revolution* (Reading, Mass.: Addison-Wesley, 1995).

44. A similar charge was leveled against the authors of *I'll Take My Stand* (see note 45) by Louis D. Rubin Jr. in his introduction to the 1962 reissue of the volume; his allegation brought swift rebuttal from some but not all of the authors. Rubin recounts the controversy in his introduction to the 1977 edition. Louis D. Rubin Jr., introduction to *I'll Take My Stand: The South and the Agrarian Tradition* by Twelve Southerners (Baton Rouge: Louisiana State University Press, 1977) (originally published in 1930).

45. Twelve Southerners, *I'll Take My Stand: The South and the Agrarian Tradition* (New York: Harper & Brothers, 1930). Two useful studies are Conkin, *Southern Agrarians,* and Mark G. Mavasi, *The Unregenerate South: The Agrarian Thought of John Crowe Ransom, Allen Tate, and Donald Davidson* (Baton Rouge: Louisiana State University Press, 1997).

46. Historian Paul Conkin views this phrase (from Donald Davidson's 1927 poetry volume *The Tall Men*) as the first use in print of the term *industrialism* by the authors of *I'll Take My Stand,* though he speculates that John Crowe Ransom had already begun to use the term. Conkin, *Southern Agrarians,* 36. According to Leo Marx, the term was originated in the previous century by Thomas Carlyle. Leo Marx, *The Machine in the Garden: Technology and the Pastoral Ideal in America* (New York: Oxford University Press, 1964), 188.

47. John Crowe Ransom, "Reconstructed but Unregenerate," in Twelve Southerners, *I'll Take My Stand* (1930), 15–16.

48. See Conkin, *Southern Agrarians,* 71, 74, 77. In his unsigned

"Introduction: A Statement of Principles," written on behalf of the twelve, Ransom did draw a parallel, claiming that "the Industrialists" were "the true Sovietists or Communists." Twelve Southerners, *I'll Take My Stand* (1977), xli.

49. Henry Kline, "William Remington: A Study in Individualism," in Twelve Southerners, *I'll Take My Stand* (1930), 313, 325.

50. Andrew Lytle, "The Hind Tit," in Twelve Southerners, *I'll Take My Stand* (1930), 203.

51. Richard Weaver, "Agrarianism in Exile," in Curtis and Thompson, *Southern Essays,* 45.

52. Leopold, *Sand County Almanac,* 200.

53. Ibid., 210.

Chapter 2: Dan Imhoff, "Linking Tables to Farms"

1. Trauger M. Groh and Steven McFadden, *Farms of Tomorrow Revisited: Community Supported Farms, Farm Supported Communities* (San Francisco: Biodynamic Farming and Gardening Association, 1997).

2. Amory Lovins, Hunter Lovins, and Marty Bender, "Energy and Agriculture," in *Meeting the Expectations of the Land: Essays in Sustainable Agriculture and Stewardship,* edited by Wes Jackson et al. (San Francisco: North Point Press, 1983).

3. Eric Sloane, *Our Vanishing Landscape* (New York: W. Funk, 1955), 14.

Chapter 4: Stephanie Mills, "Prairie University"

1. Steve Packard's "Just a Few Oddball Species" engagingly recounts his savanna species sleuthing. It is found in *Helping Nature Heal: An Introduction to Environmental Restoration,* edited by Richard Nilsen (Berkeley, Calif.: Ten Speed Press, 1991). *Helping Nature Heal* is a lively, wide-ranging, down-to-earth collection of articles on restoration projects and reviews of tools—conceptual and physical—useful to citizen-restorationists.

2. For information on Prairie University, contact the Illinois Field Office of The Nature Conservancy, 8 South Michigan Avenue, Suite 900, Chicago, IL 60603.

3. *Growing Native,* the quarterly newsletter of Growing Native, a research institute, advises home gardeners in Alta, California, about landscaping with indigenous plants. The organization's address is P.O. Box 489, Berkeley, CA 94701.

4. Laurel Ross, "The Deer Problem," *North Branch Prairie Project Brush Piles* (winter 1991–1992).

Chapter 8: Anne Mendelson, "The Decline of the Apple"

1. William Robert Prince, with William Prince, *The Pomological Manual* (New York: T. and J. Swords, 1831), vii.

2. *The Apple World: The Official Organ of the Apple Advertisers of America* 1, no. 1 (June 1914): 10.

3. Amelia Simmons, *American Cookery* (Hartford, Conn.: Hudson & Goodwin, 1796), 16.

4. Quoted in James Thacher, *The American Orchardist* (Boston: E. Collier, 1825), 14–15.

5. Ibid., ii.

6. John A. Warder, *American Pomology: Apples* (New York: Orange Judd, 1867), 14–15.

7. P[atrick] Barry, *The Fruit Garden* (New York: Scribner, 1852), iii.

8. Henry F. French, "Cultivation of Apples in the Northern States," in *Report of the Commissioner of Patents for the Year 1849* (Washington, D.C.: U.S. Patent Office, 1850), 273–76.

9. Warder, *American Pomology,* 14.

10. Horace Greeley, quoted in Sereno Todd, *The Apple Culturist* (New York: Harper & Brothers, 1871), 12–13.

11. Allen W. Dodge, "Orchards—Their Condition and Management," in *Report of the Commissioner of Patents for the Year 1849* (Washington, D.C.: U.S. Patent Office, 1850), 276.

12. Barry, *Fruit Garden,* iv.

13. A. J. Downing, *The Fruits and Fruit Trees of America* (New York: Wiley & Putnam, 1845), v–vi.

14. *The Horticulturist and Journal of Rural Art and Rural Taste* 1, no. 9 (March 1847): 310.

15. Barry, *Fruit Garden,* 278.

16. "Hints to Fruit Growers, Delivered at Oswego Institute, 1886, by Hon. Seth Fenner, of Erie County," in *Transactions of the New York State Agricultural Society* 34 (1883–1886), 200–204.

17. Dodge, "Orchards—Their Condition and Management," 277, 277 n.

18. Barry, *Fruit Garden,* v.

19. From M. L. Dunlop, "The Status of Horticulture," in *Illinois State Agricultural Society, Transactions, 1865–1866,* excerpted in Wayne D. Rasmussen, ed., *Agriculture in the United States: A Documentary History,* vol. 2 (New York: Random House, 1975), 1053.

20. L. H. Bailey, *The Apple-Tree* (New York: Macmillan, 1922), 76.

21. L. H. Bailey, *The Principles of Fruit-Growing, with Applications to Practice* (New York: Macmillan, 1915), 35.

22. Bailey, *Apple-Tree,* 77.

23. Ibid.

Chapter 11: Donald Worster, "The Wealth of Nature"

1. Lynn White Jr., "The Historical Roots of Our Ecologic Crisis," *Science* 155 (10 March 1967): 1206.

2. Karl Jaspers, *The Origin and Goal of History* (New Haven, Conn.: Yale University Press, 1959), 1–21.

3. René Descartes, "Discourse on the Method," in *The Philosophical Writings of René Descartes,* trans. John Cottingham, Robert Stoothoff, and Dugald Murdoch (1637; reprint, Cambridge, England: Cambridge University Press, 1985), i, 142–43.

4. Adam Smith, *An Inquiry into the Nature and Causes of the Wealth of Nations,* edited by Edwin Cannan (1776; reprint, New York: Modern Library, 1937).

5. This passage is from Locke's *Some Considerations of the Consequences*

of the Lowering of Interest and Raising the Value of Money (published in 1696), quoted in Smith, *An Inquiry,* lvii.

Chapter 13: Brian Donahue, "Reclaiming the Commons"

1. Henry David Thoreau, "Huckleberries," in *The Natural History Essays* (Salt Lake City: Peregrine Smith, 1980), 255–56.

Acknowledgments

Donald Worster was kind enough to review my draft introduction and to offer comments that stimulated many good changes. I thank him for doing so. My thanks also go to Jonathan Cobb, executive editor of Shearwater Books, who helped guide the project from its inception and was available at every step to respond to ideas. I'm particularly grateful for his willingness to let me know, without equivocation, when I strayed from a sensible path.

The four individuals to whom the book is dedicated—students, colleagues, friends—helped and inspired me more than they realize. I here record, with pleasure, my gratitude to them as well as my hope of many good dealings in years to come.

"Learning from the Prairie" is from *The Force of Spirit* by Scott Russell Sanders. Copyright © 2000 by Scott Russell Sanders. Reprinted by permission of Beacon Press, Boston.

"Linking Tables to Farms," copyright © 1999, 2001 by Dan Imhoff. Used by permission of the author.

"Substance Abuse" is from *This Place on Earth: Home and the Practice of Permanence* by Alan Thein Durning. Copyright © 1996 by Alan Thein Durning. Reprinted by permission of Sasquatch Books, Seattle, Washington.

"Prairie University" is from *In Service of the Wild* by Stephanie Mills. Copyright © 1995 by Stephanie Mills. Reprinted by permission of Beacon Press, Boston.

"The Whole Horse," copyright © 1999, 2001 by Wendell Berry. Used by permission of the author.

About the Contributors

WENDELL BERRY works a farm in Henry County, Kentucky, with his wife, Tanya, not far from where he was born. Hailed as one of America's leading writers and moralists, he is the author of more than three dozen books of essays, fiction, and poetry, including *Jayber Crow, A Timbered Choir, What Are People For?* and *The Gift of Good Land.* His work has earned numerous fellowships and awards, including the T. S. Eliot Prize, the John Hay Award, the Lyndhurst Prize, and the Aiken-Taylor Award for Modern American Poetry.

BRIAN DONAHUE is chair of the Environmental Studies Program at Brandeis University. He cofounded and for twelve years directed Land's Sake, a nonprofit community farm in Weston, Massachusetts. For three years, he was director of education at The Land Institute, near Salina, Kansas. He is the author of *Reclaiming the Commons: Community Farms and Forests in a New England Town.*

ALAN THEIN DURNING is founder and executive director of Northwest Environment Watch in Seattle. A former senior researcher at Worldwatch Institute, he has written widely for publications such as the *New York Times, Foreign Policy,* and the *Washington Post.* His books include *How Much Is Enough?* and *This Place on Earth: Home and the Practice of Permanence.*

ERIC T. FREYFOGLE is the author of *Justice and the Earth* and

Bounded People, Boundless Lands, recipient of the 1999 Adult Non-fiction Award of the Society of Midland Authors. His forthcoming book is *Land Matters: Private Property, Politics, and the Common Good.* A native of central Illinois and an active local conservation-ist, he teaches property, environmental, and natural resources law at the University of Illinois College of Law, where he is Max L. Rowe Professor.

DAN IMHOFF is a writer, publisher, and homestead farmer living in northern California's Anderson Valley.

WILLIAM KITTREDGE, of Missoula, Montana, has written exten-sively about the links between people and the working landscapes of the American West, particularly the ranchlands where for many years he lived and worked. His books include *Balancing Water, Hole in the Sky,* and *Who Owns the West?* He served as editor of *The Portable Western Reader* and coproduced the film *A River Runs through It.* His honors include the National Medal for the Human-ities.

DON KURTZ grew up in the farm country of east-central Illinois before moving to the Southwest. He is the recipient of a National Endowment for the Arts Creative Writing Fellowship, and his work has appeared in a wide variety of magazines. Now on the fac-ulty of New Mexico State University, he is the author of *South of the Big Four.*

DAVID KLINE operates a certified organic dairy farm in Holmes County, Ohio, with his wife, Elsie, and their family. A regular con-tributor to Amish publications, he is the author of *Great Possessions: An Amish Farmer's Journal* and *Scratching the Woodchuck: Nature on an Amish Farm.* He is cofounding editor, with his wife, of *Farming Magazine: People, Land, and Community.*

GENE LOGSDON and his wife, Carol, farm thirty-two acres in Upper Sandusky, Ohio, a mile from his boyhood home. He is the author of hundreds of essays for such publications as *Orion, Whole Earth Review,* the *Utne Reader, Organic Gardening,* and *Draft Horse Journal.* His many books on small-scale farming and organic gardening include *Living at Nature's Pace, Good Spirits, The Contrary Farmer,* and *The Contrary Farmer's Invitation to Gardening.*

ANNE MENDELSON, of New Jersey, is a food historian and longtime writer on food and cooking. She is the author of *Stand Facing the Stove: The Story of the Women Who Gave America the Joy of Cooking* and coauthor of a forthcoming work, *Zarela's Veracruz.*

STEPHANIE MILLS, who lives near Maple City, Michigan, has been for three decades a prolific writer and speaker on issues of ecology and social change. Her writings include *In Service of the Wild: Restoring and Reinhabiting Damaged Land, Whatever Happened to Ecology?* and a forthcoming book, *Epicurean Simplicity.*

DAVID W. ORR chairs the Environmental Studies Program at Oberlin College in Ohio, where he has gained worldwide attention for leading the design of the ecologically sensitive Adam Joseph Lewis Center for Environmental Studies. A trustee of many foundations, he has received many recognitions, including the National Conservation Achievement Award of the National Wildlife Federation, the Lyndhurst Prize, and an honorary doctorate in humane letters from Arkansas College. He is the author of *Ecological Literacy, Earth in Mind,* and a forthcoming book, *The Nature of Design.*

SCOTT RUSSELL SANDERS is Distinguished Professor of English at Indiana University, Bloomington, where he heads the prestigious Wells Scholars Program. Many of his books, including *The Force of Spirit, Hunting for Hope, Writing from the Center,* and *Staying Put,* display his longstanding fascination with farms, gardens, landscapes, and the people who inhabit them. He has received numerous literary awards, and his essays have been widely reprinted.

DONALD WORSTER has for several decades been one of the nation's leading environmental historians. He lives with his wife, Bev, who raises sheep, in his home state of Kansas. For the past decade he has been Hall Distinguished Professor of American History at the University of Kansas. His many books include *A River Running West, An Unsettled Country, The Wealth of Nature,* and *Dust Bowl,* winner of the Bancroft Prize in American History.

Index

Accumulation, 154. *See also* Property ownership values

Aesthetics, moral or natural, xv, xxxiii, 104

Agrarianism: as a land-based culture, xiii, 93, 97, 148; moral code of, xxx–xxxi; vs. industrialism, 63 (*Ch.* 5), 67–68, 70–71, 84, 94, 99. *See also* New agrarianism; Traditional agrarianism

Agrarian mind, xiii–xli passim, 68, 69–70, 101, 152–53; fictionally depicted, 239 (*Ch.* 15); the urban-agrarian mind, xxxvii, 93 (*Ch.* 7), 106–7

Agribusiness. *See* Industrial agriculture

Agricultural economics, 82–83, 84

Agriculture, 114, 147–48; changes over time in, 82–84. *See also* Farming; Industrial agriculture

Agriculture, U.S. Dept. of, 10, 14

Ahistoricism, 161; industrialism and, 64–65. *See also* Historicism

Alternative energy technologies, 215; bio-diesel fuel, 12; photovoltaic cell arrays, 12, 103

Alternative food production, xvi, 87–90

American Farm Bureau Federation, xxxv

American Fruit Culturist, 115

American West: frontier myths, xxxv, 145, 152–54; frontier period, xxxv, 36–37

Amish farm, 13, 97, 102, 181 (*Ch.* 12); community life on, 193–95; family life, 181, 182–85; farming methods, 186–92

Animal rights activism, 55–56

Animal wildlife, 48

Animism, 166

Annual crops, 7–8

Apple Advertisers of America, 112–13

Apples: commercial orchards, 119–25; decline of the apple, 111 (*Ch.* 8), 122–26; early varieties and specialized growing of, 111, 119–23; global marketing of, 124–25; new and "heirloom" varieties of, 121–23, 127. *See also* Orchards

Aristotle, xxix

Bacon, Francis, 170

Bailey, Liberty Hyde, 126, 127

Balaban, John and Jane, 45–46, 47, 49

Barry, Patrick, *The Fruit Garden,* 119, 122, 123–24

Barsotti, Kathy, 18

Barter exchange, xxiv